THE MATING GAME

ROBERT BURTON

THE MATING GAME

Crown Publishers, Inc., New York

Editors: PETER HUTCHINSON PHD
HOWARD LOXTON
Managing Editor: BEN LENTHALL
Picture Editor: POLLY FRIEDHOFF
Design: ELWYN BLACKER
Production: ANDREW IVETT
Drawings: J. R. FULLER

Planned and produced by ELSEVIER INTERNATIONAL PROJECTS LTD, OXFORD

First published in U.S.A. 1976
by Crown Publishers, Inc.

Library of Congress Cataloging in Publication Data

Burton, Robert, 1941-
The mating game.

1. Sexual behavior in animals. 2. Reproduction.
I. Title.
QL761.B86 591.5'6 76-7998
ISBN 0-517-52632-8

Origination by ART COLOR OFFSET, ROME, ITALY
Filmset by KEYSPOOLS LIMITED, GOLBORNE, LANCASHIRE, ENGLAND
Printed and bound
by Brepols, Turnhout, Belgium

TITLE PAGE: Mayflies swarming over water. Most of their lives are spent as aquatic larvae. The adults exist for one purpose. They breed and then die.

ACKNOWLEDGMENTS

Artists and photographers are listed alphabetically with their agent's initials, where applicable, abbreviated as follows:

Bruce Coleman Limited (Res)
Frank W. Lane (FWL)
Natural History Photographic Agency (NHPA)
Natural Science Photos (NSP)

A. Bannister (NHPA) 10 (bottom), 40 (left), 56, 60, 70, 72, 78, 79 (top), 95
A. C. Bisserot (Res) 53
F. V. Blackburn (NHPA) 102
J. B. Blossom (NHPA) 113, 114, 115 (centre)
R. & M. Borland (Res) 146 (top), 146 (bottom)
M. Boulard (NSP) 76
J. Burton (Res) 12 (left) 12 (right), 13, 14, 15, 16, 18 (bottom), 19, 24 (top right) 24 (top left), 26, 27, 28, 29 (top) 29 (bottom), 31 (top) 31 (bottom left), 32, 34, 35, 36, 37 (top), 39, 41, 43, 45, 50, 60 (top) 60 (bottom), 61, 62, 63, 64, 66, 71, 74, 77, 85, 86 (left), 88, 90, 92, 94, 100, 111, 122 (top), 128, 130 (top), 134, 137, 138, 151, 152
R. Burton, (Res) 110 (bottom)
A. Christiansen (FWL) 109
M. Coe, 106 (top)
M. J. Coe (NSP) 96 (bottom)
G. Cubitt (Res) 143
S. Dalton (NHPA) 38 (top)
P. M. David (Seaphot) 10 (top)
Ekdotike Athenon S. A. Athens 8
D. Fisher (FWL) 46
J. B. Free 59
J. Fuller 33, 40 (right), 47, 57, 67, 79 (bottom) 86 (right), 103, 139
F. Greenaway (NHPA) 97
C. M. Hladik, 142, 145 (top) 148 (left) 148 (right), 149
Jacana 75, 104, 108, 122 (bottom) 124 (top) 124 (bottom), 125, 127 (top), 136
P. Johnson (NHPA) 121 (top)
J. Karmali (FWL) 112 (top), 127 (bottom)
F. W. Lane (FWL) 107, 119, 126, 130 (bottom)
L. Lyon (Res) 150
G. Mazza 55 (top), 55 (bottom), 98, 101, 117, 121 (bottom)
G. J. H. Moon (FWL) 115 (top)
A. E. Mourant 155
N. Myers (Res) 144
A. van den Nieuwenhuizen 6, 25, 96 (top), 99
Oxford Scientific Films title page, 17 (top), 24 (bottom), 30, 42 (top), 42 (bottom), 48, 49, 51, 58 (top), 58 (bottom), 87 (top), 87 (bottom left), 87 (bottom right), 120, 132, 145 (bottom)
L. E. Perkins (NSP) 89
G. Pizzey (NHPA) 105 (bottom), 123 (top)
I. Polunin (NHPA) 31 (bottom right)
K. G. Preston-Mafham (NHPA) 52 (top), 52 (bottom), 69, 83
S. Price (NSP) 123 (bottom)
M. Quarishy (Res) 147
A. Saunders (FWL) 110 (top)
Spectrum Colour Library 73, 106 (bottom), 129, 140 (top), 140 (bottom), 156
R. Thompson (FWL) 91
Tony Stone Associates 133
M. Walker (NHPA) 17 (bottom), 18 (top), 20, 21, 22
M. I. Walker (Philip Harris Biological) 11
P. H. Ward (NSP) 38 (bottom), 44, 68
C. E. Williams (NSP) 80, 81
F. H. Wylie (FWL) 37 (bottom)
G. Zeisler 105 (top), 112 (bottom), 115 (bottom), 116, 131

The Publishers have attempted to observe the legal requirements with respect to the rights of the suppliers of photographic materials. Nevertheless, persons who have claims are invited to apply to the Publishers.

CONTENTS

INTRODUCTION

Animals and plants are unique in that they can reproduce themselves and so ensure survival beyond the life-span of any individual. We do not yet fully understand why animals and plants grow old, but they do, and it is as important for a single celled animal that may only live for days or even hours to reproduce its own kind as it is for a redwood fir which, under favourable conditions, may live for well over a thousand years.

Some animals reproduce by a process of simple division that results in the production of offspring identical in every respect to the parent. Most, however, reproduce sexually. This book is mainly concerned with aspects of the sex life of animals. Sexual reproduction differs in one extremely significant way from asexual reproduction. In sexual reproduction the genetic material that largely determined the characteristics of individual offspring is contributed by both parents. Which genetic material comes from the mother, and which is contributed by the father is the result of partly random events that affect the production of eggs and sperm. As a result, offspring differ not only from each other, but also from their parents. The function of sex is thus to ensure that animal populations are composed of a variety of individuals. As Darwin demonstrated over a hundred years ago, it is this variation that enables animals to change from one generation to the next in order to take advantage of new habitats and to keep pace with the ever changing environment.

Genetically, sexual reproduction is essentially the same for all animals and this aspect is described in the early chapters. The author then goes on to examine the incredibly varied ways in which different animals reproduce successfully. In many 'lower' animals the fertilization of an egg by a sperm takes place in the open sea. To ensure that this happens, such animals are sometimes forced to produce enormous quantities of eggs and sperm. In some groups, physiological and behavioural mechanisms have evolved that bring members of each sex together during the mating season. These make the production of fewer eggs and sperm possible by enhancing the chances of fertilization. Other groups have eliminated almost every effect of chance by evolving the ability to fertilize eggs inside the female's body. This extremely successful strategy can only work if males and females of the same species can locate one another and overcome their instincts, developed for entirely different reasons, to either eat or flee from any other creature. Thus some of the most fascinating aspects of animals have evolved: territorial behaviour; courtship ritual; monogamy; polygamy and the mechanics of mating itself.

Two Apple snails, or Edible snails, mating.

THE BEGINNINGS OF SEX

In a society where individuals can control their fertility by contraception or abstention, we tend to forget how powerful is the natural drive to beget offspring. It was otherwise in pagan Europe, where the continued fertility of Man (used here with a capital M in the sense of *Homo sapiens*, both male and female of the species), his livestock and his crops was an overwhelming concern. Fruitfulness was so important to primitive agricultural people that their religions were largely aimed at ensuring fertility in all things and remnants of their beliefs have survived in folk memory to the present day. The universal importance of continued fertility was embodied in the idea of the great Earth Mother, who controlled the birth of animals and the germination and fruitfulness of plants. Our ancestors saw her power in the quickening of life every spring, when plants spread their leaves and petals, and animals conceived and bore their young.

In their mythology our early forebears held Woman to be supreme. Society was matriarchal; there was no Sky Father, in the form of Zeus or Odin, and the Earth Mother's consort played only a limited role. His sole function was to fertilize and, in many guises throughout Europe and the Near East, there was a springtime ceremony to mark the accession of the Earth Mother's consort. His short summer reign was followed by his banishment in autumn and his return in spring.

The joyous reunion of the Royal couple to promote spring fertility continued to be celebrated, if only subconsciously, throughout the Christian era and is still preserved in May Day parades, spring weddings and Easter eggs. The survival of these spring rites was no doubt assisted by the Christian doctrine carrying the same message.

The importance of reproduction is emphasized in the first chapters of the Book of Genesis. Here there is a definite command: 'Be fruitful, and multiply, and replenish the earth'. Man and animals and plants are commanded to reproduce their kind and populate the earth's surface. There seems also to be a clear indication of how reproduction is to be effected. There is no mention of sexual reproduction as such but the story of the Garden of Eden does deal with the origin of the sexes; and the union of male and female to produce offspring is implicit in the animals entering the Ark 'two by two'.

The patriarchs and scholars responsible for setting down the Bible story were familiar with the reproductive habits of their own kind and of their domestic animals. They had also watched the courtship of birds, the nest making and the laying of eggs, and they could not possibly have missed the mating of their domestic animals. In other words, they were familiar with what we now call the facts of life.

These facts were quite definite. To be fruitful and multiply, both male and female are needed, the male to fertilize the female; in human terms this means having first wooed her and gained her acceptance. But the labour of reproduction – 'In sorrow thou shalt bring forth children' – is entirely the female's, although the male may help care for the family.

Such was the immutable law of life from which escape was impossible; but the law was obeyed blindly. The secrets of human reproduction had not been unlocked nor the implications fully appreciated. Now, thousands of years later, our knowledge, as well as our experience, are so much wider. We have learned to investigate the natural world, to enquire about what we find, and to question the old beliefs. Yet the original command to multiply and replenish the earth still holds for man, beast and plant and the study of

The Earth Mother, the symbol of fertility and senior deity for primitive religions the world over. The male principle was subordinate to her.

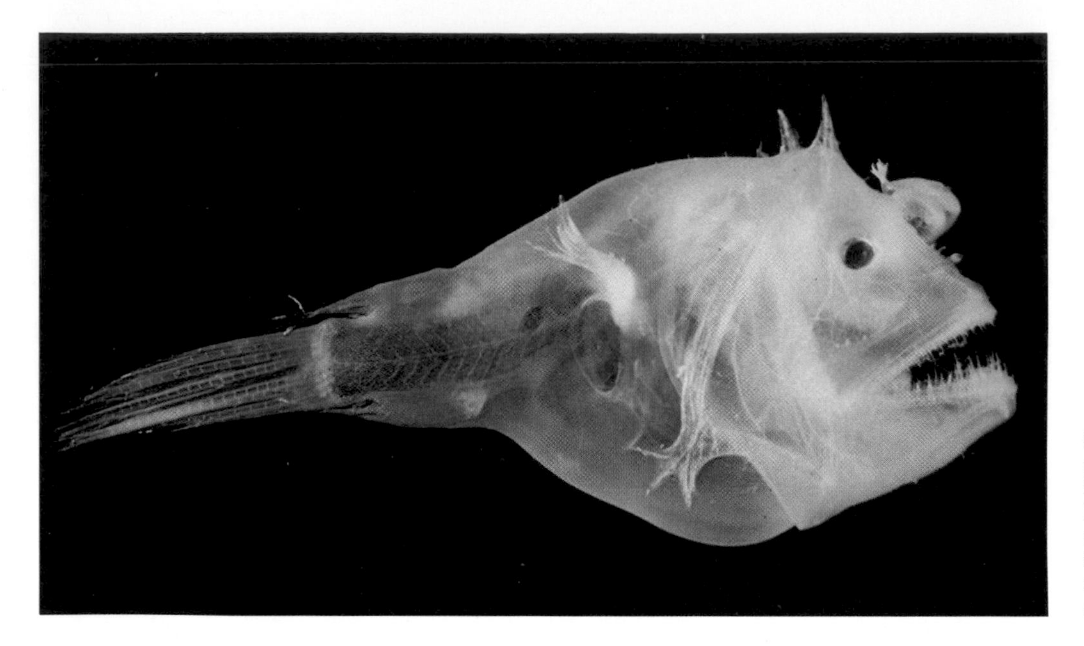

Subservient males of the Deep-sea angler fish live parasitically on the female. Once a male Deep-sea angler has found a female he fastens to her and degenerates until he is no more than a bag of reproductive tissues, nourished by the female and having the sole function of fertilizing her eggs.

biology has revealed its full force. Each species of animal and plant is endowed with an overwhelming drive to procreate and to fill, if not the earth, that part of it to which it is suited, its habitat. Moreover, every individual is seen to exist solely for the purpose of procreation. Each species can be likened to a totalitarian state in which the individual is unimportant and survives solely for the benefit of the whole.

Few animals survive long beyond the period of reproductive usefulness. Man is virtually the only animal to survive after his reproductive days are over and is the only species to limit consciously his reproductive capacity. He questions the command that his first purpose is to procreate. At the present time there are two lines of argument for disobeying the command: the human species as a whole has got out of hand and we are dangerously overpopulating the earth, while at the individual level there is a growing resentment by women that their main, almost their only, function in some cultures should be to bear children.

For the moment we must leave the unique

The symbol of the dominant female, the Praying mantis eats her mate. His sacrifice is not futile; his body furnishes nourishment for the 100 or so eggs, seen here being laid in a white frothy case.

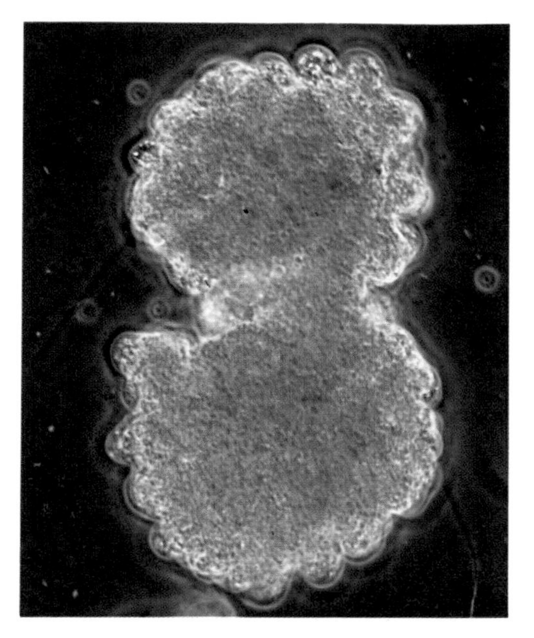

Amoeba shows sex at its simplest. The organism merely splits in two. The two 'daughters' are identical to each other and to their 'mother'.

position of Man and consider only the remainder of the animal kingdom where the command to procreate continues unabated. Indeed, only by leaving Man out of the discussion, and forgetting morals and principles, can we look dispassionately at the problems of reproduction, although it will become clear that there is no basic biological difference between mating in Man and in other animals. We can trace the development of sexuality by looking at its manifestation in the various groups of animals in ascending order of organization.

It is not necessary to make too detailed a survey to appreciate that the methods of reproduction used by animals are many and varied. There are almost as many ways of courting, mating and producing young as there are animal species and some animals have quite bizarre reproductive habits. Take the slugs that mate while each is dangling from its slimy rope, or the snails that cannot mate without wounding each other with 'love darts'. Among higher animals are the dogs that become locked together during mating and the bull Grey seals that do not take 'No' for an answer.

Perhaps the most remarkable of all the bizarre matings known concerns deep-sea anglerfishes and the Praying mantis, a species of insect in which the female eats her mate during copulation. In the anglerfishes the male is fused with the female from his early days, becoming permanently and utterly dependent on her throughout the rest of his life, his whole body degenerating so that he is no more than a bag containing the reproductive organs. The male Praying mantis often has his head bitten off during mating but this, if anything, facilititates the act because it removes the gland secreting the hormones that inhibit his mating impulse. Both of these will be discussed later.

Nature goes to extreme lengths, as it were, to carry out a basic function and it is our aim here to examine the reasons for different kinds of reproductive behaviour. Reproduction without sex – asexual reproduction – was undoubtedly the way the first animals multiplied and for some it is still adequate. For them, sexual reproduction would seem to be an unnecessarily elaborate procedure. Yet there must have been good reason why the majority of animals have developed the sexual method. So it would seem that both asexual and sexual reproduction have their own advantages and disadvantages.

The extraordinary lengths to which some animals go to 'have sex' suggests that their strange sex lives would not have evolved without some very strong pressure to do so.

Reproduction without sex. A story current a few years ago concerned a young woman at a cocktail party who, having listened to a discussion on the mysteries of creation, asked naively: 'Do you think sex is here to stay?' When one comes to contemplate the matter, her question is not so silly as might first appear. In many areas of the animal kingdom reproduction is asexual, or, as some authors prefer, non-sexual; virgin birth is not wholly unknown; and even though non-sexual reproduction is found mainly among the lower animals there is at least one vertebrate, a species of lizard living in the Caucasus, for which no males have so far been found.

The first of these loveless species, the one usually taken to typify the simplest method of reproduction, is the unicellular animal amoeba which figures in every textbook in biology and which is familiar to all students of that science. *Amoeba proteus*, to give it its full scientific name, is a common inhabitant of ponds, and is famous as the accepted prototype of living organisms used to typify, quite wrongly, the ultimate ancestor from which all other animals have descended. It is to all intents a simple blob of protoplasm enclosed by a membrane, a single living cell, lacking skeleton, sense organs, brain and nervous system, moving by throwing out armlike extensions of its body called pseudopodia and feeding by engulfing smaller organisms. In amoeba all the functions of life are controlled by a central nucleus and, like all other living things, the amoeba is able to reproduce itself.

At some point in its life, the amoeba undergoes certain changes as a result of which it gradually splits into two. The first stage in this simple binary fission, as it is called, is that the amoeba becomes spherical. Its nucleus then divides into two and the two halves pull apart. Meanwhile, the cell itself begins to split into two and when this is complete there are two new 'daughter' amoebae, each with its own nucleus. The whole process takes less than an hour from beginning to end after which the daughter amoebae feed and grow. When they have reached a certain size each divides again. So fission of amoeba cells continues unabated so long as food is plentiful, and a clone is formed. A clone is a colony of identical animals.

The remarkable thing is that although the amoeba has been watched by literally thousands of scientists for a good two centuries, no one has ever seen it reproduce other than by simple binary fission. Sex is completely missing from its life so it would seem

A starfish can replace an arm lost by accident and it can reproduce by splitting and regeneration. The animal literally pulls itself in two. Each part grows the missing pieces. This series shows the value of regeneration. A Painted coral prawn or Harlequin shrimp is feeding on starfish watched by three Golden seahorses (BELOW, LEFT). The prawn picks away with special pincers until the arm is severed and the starfish may even assist by throwing off the arm (BELOW). The starfish has lost one arm but the remainder survives. Regeneration starts immediately (RIGHT) and tube feet grow on the stump as soon as the wound has healed.

that sexual reproduction is not, as has often been asserted, an essential part of life. This is supported by what is found in animals with a considerably more advanced level of bodily organization.

Take starfish, for example, which are male and female and, normally, reproduce sexually. On occasion their behaviour appears to become deranged, for instead of their five arms working in unison to propel the starfish over the seabed, two start to move in the opposite direction to the remaining three, and the starfish splits in two. Each of the two parts, consisting of half a body, one part with two arms, the other with three, goes its own way and grows the missing portion of the body and the extra arms so that each ends up as a complete starfish.

Dr Mortensen, the distinguished Danish zoologist who spent his lifetime studying the echinoderms (starfishes, Sea urchins and their relatives), has recorded how he investigated a reef in the Caribbean which was populated by numerous brittlestars, close relatives of starfishes. He found that they were all males. His assumption was that a male brittlestar originally found its way to the reef, divided by splitting in two and

The Snakelocks anemone can reproduce by splitting in two. The two halves gradually draw apart until suddenly there are two anemones where before there was one.

LEFT: Beadlet anemone with babies. They develop from buds inside the anemone's body and are expelled through its mouth.

thereafter the only method of reproduction possible in this isolated colony was fission. This striking example makes clearer an idea often expressed by zoologists of the early years of this century, that the amoeba has found true immortality. The same can be said of Mortsensen's colony of brittlestars, for the original male that split on the reef was still alive in all the others populating the reef when Mortensen found it.

There is another way in which animals can reproduce asexually. This is by budding, a process normally associated with plants but also widespread among the lower animals. For example, a huge coral weighing half a ton may be the product of a single larva which by a process of continual budding has produced the living polyps that have laid down the chalky part of the coral and live within it. The process of coral formation can be best understood by reference to the simpler situation found in another textbook creature living in ponds, known as *Hydra*. This was named after the mythical creature encountered by Hercules, the many-headed Hydra, which grew fresh heads as fast as Hercules cut them off. The modern *Hydra* is a polyp related to Sea anemones whose body is a simple cylindrical bag, with a two-layered wall, fixed at one end to a water-plant. At the other end is a ring of tentacles which catch Water fleas and other small animals and push them into a mouth that lies in the middle of the ring. *Hydra* can reproduce both asexually and sexually. Its asexual reproduction is, however, not the simple binary fission of the ancestral amoeba. It is by budding, as in corals. In this, a small area of tissue on the side of the body starts to grow rapidly. That is, each cell in this area divides repeatedly and rapidly and a bud is formed, looking like a wart on the side of the *Hydra*. This lengthens and changes in shape

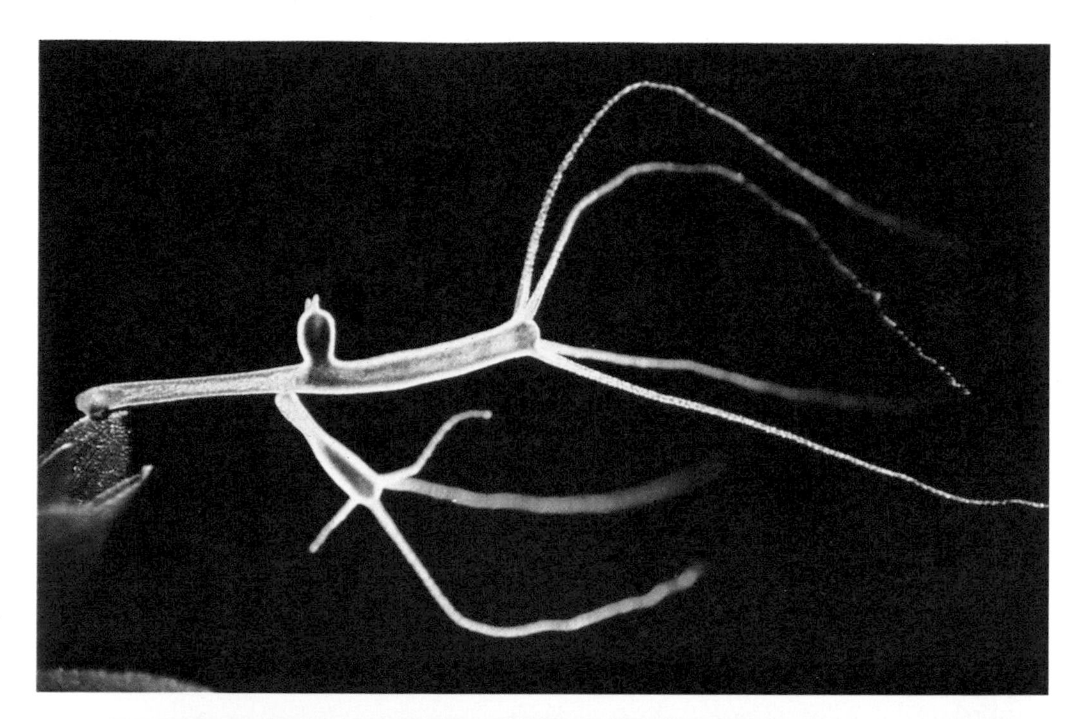

Simple sex in *Hydra*. When food becomes scarce, the animal develops testes and ovaries on the outside of the body. The testes are shown here, from which the sperms swim through the water to fertilize the single ovum.

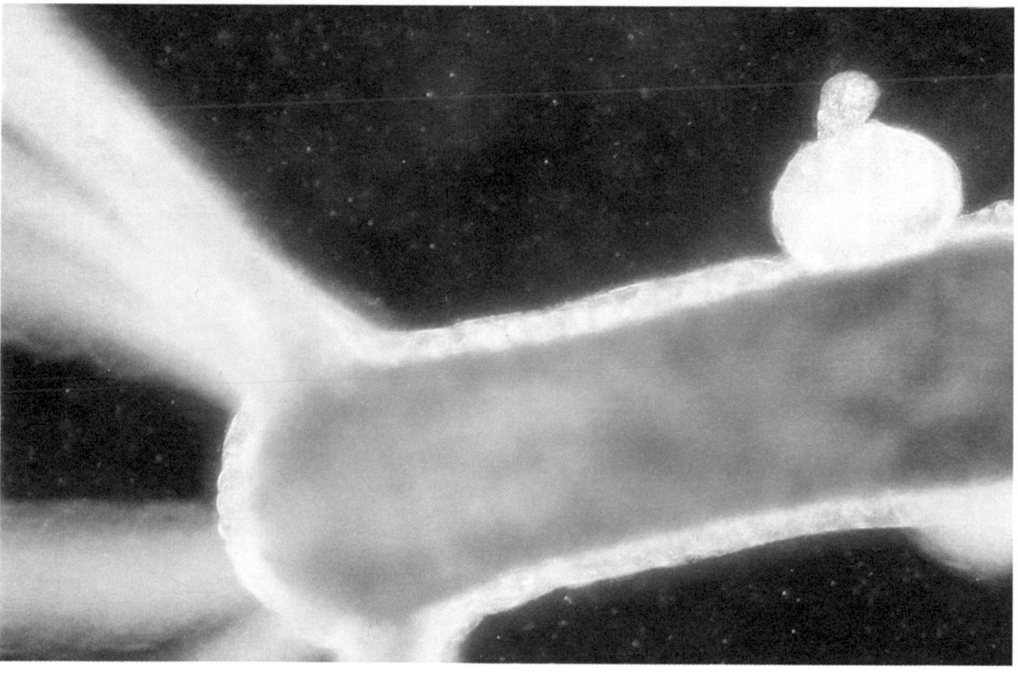

Hydra, a freshwater relative of the sea anemones, reproduces by budding when food is plentiful. A newly-formed bud is seen on the upperside. This will grow tentacles and start to feed itself. The lower bud is almost ready to detach itself and lead an independent life.

LEFT: Sea squirts *Botryllus* are commonly found growing on seaweeds. The delicately patterned colony results from repeated asexual budding. The individuals are buried in a tough jelly so that they can survive the pounding of waves.

until a minute polyp is formed complete with tentacles. After a while an opening, the mouth, appears at the free end, the polyp begins to feed and it then seals itself off at the base and floats away to settle down and lead an independent life.

Some marine sponges use both these methods. That is they reproduce both by tearing themselves apart and by budding, and they use these methods in addition to producing eggs and sperm and therefore swimming larvae. It is not generally known that there are 2,500 different kinds of sponge and those used commercially and for toilet purposes represent only a handful of species. Among the known commercial sponges there are several known to reproduce in this surprising combination of ways. One is the Purse sponge that grows on the lower shore on rocky coasts throughout much of the world. It is always pictured as having the outline of a brandy glass but flattened from side to side and white or cream coloured. This shape, in its perfect form, is achieved only by the young Purse sponge, because as it grows there appear across the body lines of weakness, the body becomes folded in various ways or even perforated, buds are nipped off at the margins, and there comes a time when all the Purse sponges over a wide area are likely to break up into fragments. Each fragment reforms itself so that it has the typical brandy glass shape and attaches itself by its stalk to a solid support. It would be difficult to say whether

LEFT: A ripe ovum of *Hydra* being extruded from the body. It will be fertilized by a sperm to produce a new individual.

more Purse sponges are produced by fragmentation or budding or more by the production of larvae. There is one certainty, that the Purse sponge has more than one string to its bow. This may be why it is so widespread and numerous.

The Purse sponge is not alone among marine sponges in reproducing by fragmentation but it is the most spectacular. Purse sponges hang down from the undersurfaces of rocks or from the walls and ceilings of undersea caves. Their average height is about two inches but they may be up to five inches and their pale colouring makes them conspicuous. The fragmentation seems to take place suddenly and simultaneously throughout a given locality. So it is possible to inspect an undersea cave one day in spring and find its walls and ceilings decorated with hundreds of these sponges, all apparently intact. A few days later there will be a scene of destruction as if somebody had walked round the cave systematically slashing at the sponges with a stick in order deliberately to break them up. Tattered remains hang from the roof and the walls and broken fragments

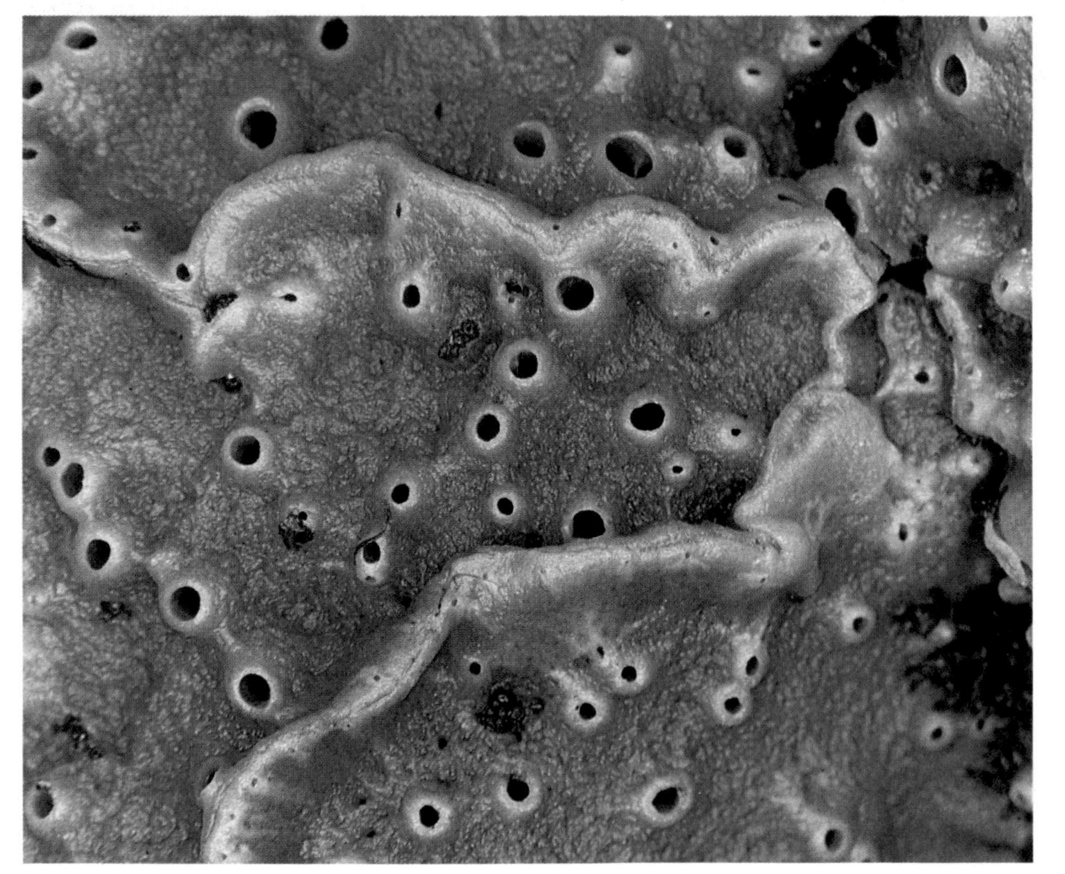

RIGHT: Only in the 19th century did scientists decide that sponges, like these Breadcrumb sponges, are animals and not plants. They reproduce both sexually and asexually.

LEFT: The top left-hand individual of these Purse sponges is beginning to divide into two, one of several ways in which these sponges reproduce asexually.

are scattered around. Presumably, something has happened akin to the spawning crisis which will be mentioned later.

Having examined the reproductive habits of these simple marine and freshwater animals, two questions now present themselves. The first is: How did the differentiation into sexes come about? The other is: Why, if asexual reproduction can achieve the birth of offspring, was it necessary for animals to turn to the more complicated sexual method?

There is relatively little direct evidence about the beginnings of sex in the animal kingdom. Such as we have does enable us to picture how it could have come about.

Simple sex. The same pond water that yields amoebae is also the home of several other kinds of protozoan. One, descriptively called the Slipper animalcule but now referred to by its scientific name *Paramoecium*, grows well in an aquarium culture, rapidly multiplying by binary fission every eight hours. Usually

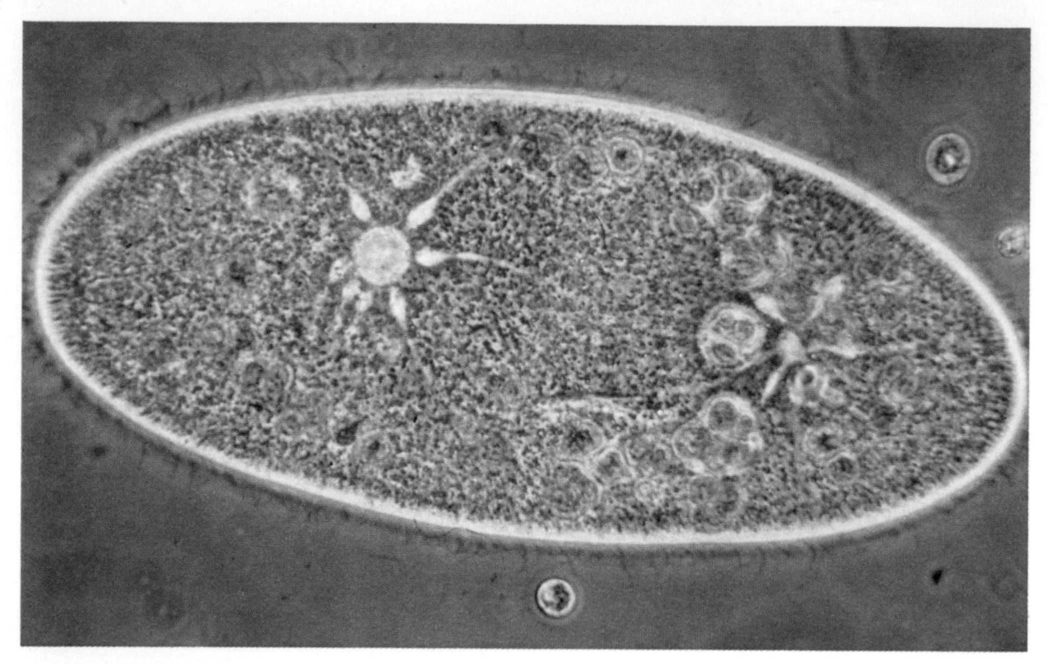

Paramoecium is a single-celled animal that can reproduce with or without sex. It spins through still fresh waters collecting bacteria and other minute organisms in its gullet.

Paramoecium behaves in the way already described for amoeba. If, however, a new batch or clone is brought in from a pond and placed in an aquarium in which another clone of *Paramoecium* is living, this 'wild' clone immediately pairs up with the 'domestic' clone in the aquarium in what is known as conjugation. Individuals from each clone come together in pairs and fuse. The adjacent cell membranes of each pair break down and they exchange parts of their nuclei. This is a complicated process because protozoans of the ciliate group, to which *Paramoecium* belongs, possess three nuclei, the macronucleus and two micronuclei. At conjugation the macronucleus in each *Paramoecium* disappears and each micronucleus divides into four. Seven of the daughter micronuclei disappear while the eighth divides again. Then one micronucleus from each *Paramoecium* passes into the body of the other and fuses with the micronucleus that has remained in the original cell. The cells then separate and each divides into four daughter cells. The paramoecia now go their own way and continue to reproduce by simple binary fission.

The important thing to note is that the partners in the pairing of *Paramoecium* are equal. Thus there is the appearance of a sexual union but with complete equality of the sexes. This is why it is preferable to speak of this process as conjugation instead of sexual union. Conjugation does, however, foreshadow what takes place in true sexual union. In other protozoans this is taken further. *Chlamydomonas* is another common pond protozoan. In some species of *Chlamydomonas* the two members of a conjugating pair are identical as in *Paramoecium* but in others one of the pair is larger and less mobile and is the forerunner of an egg or ovum. The other is smaller, more active and is the equivalent of a sperm. That is, there are the beginnings of a differentiation into separate sexes, giving what is known as sexual dimorphism.

It is now possible to speak of the conjugating pair as male and female. The small active male seeks out the larger, less active, female. This is, by definition, the basic difference between male and female. Whether we are dealing with the mobility of sperms seeking the ovum or the initiative taken by a male animal courting the female, as in the higher animals, it is the male that is typically the more active and the female that is typically the more passive, the female having forfeited mobility in favour of a large food store needed for the development of the embryo. As we pass up the animal scale, there is a progressive increase in sexual dimorphism from the slightly different partners seen in protozoan conjugation to the extreme forms of sexual differences that reach their peak in the showy plumes and extravagant displays of birds or the 4:1 size difference between male and female Elephant seals and sealions.

The evolutionary progress of sexual dimorphism is not steady and cannot be expected to be so, because nowhere in the living world is there a tidy pattern of change. There are gaps, duplications and back-tracking which make it impossible to describe a neat story of change from the most primitive to the most advanced condition. We shall see several times how one group of animals has evolved several ways of reproduction and how, on the other hand, the same particular method of reproduction can be found in widely separated realms of the animal world.

One of the most difficult gaps to bridge in the sequence of animal evolution is that between the unicellular protozoans and the multicellular animals or metazoans. How the

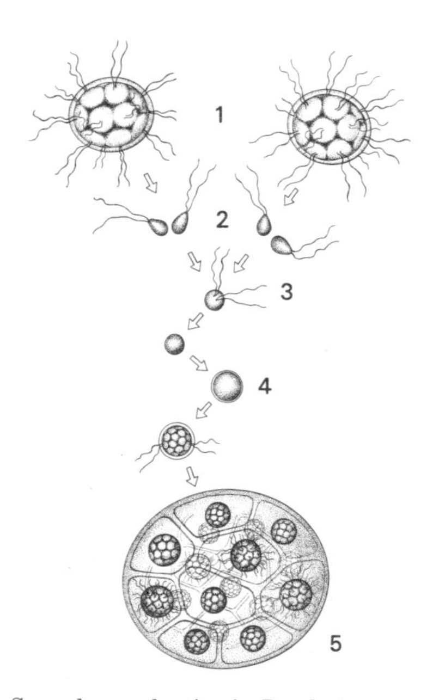

Sexual reproduction in *Pandorina*. Cell colonies of 16 to 32 cells (1) release zooids (2) which conjugate or fuse (3) to produce a zygote (4). This grows and divides to produce a new colony (5).

Sometimes reproduction in *Paramoecium* is sexual; two individuals lie side by side, joining in the region of the gullet. Genetic material is exchanged, the two break apart then each divides into new 'daughter' individuals. More frequently, reproduction is by simple division. The centre of the body gradually constricts as if a fine loop of thread is being pulled tight, and the halves split apart.

latter arose from the former is not known for certain but it seems likely that in some protozoan, dividing by binary fission, the parts failed to separate. At the next division the cells still failed to separate, so a whole colony of united cells was formed. This mass of cells set the stage for a division of labour among the cells. Some became primarily concerned with locomotion while others became sense organs or nerves, or specialized in feeding. And a few became germ-cells from which new colonies were formed.

Something approaching this level of organization is shown by yet more text-book pond organisms which are classed as plants because they make their own food by photosynthesis although they have retained the animal characteristic of being able to move about. Such organisms have in the past been referred to as plant-animals. They are important in illustrating a possible way in which multicellular organisms may have developed from unicellular organisms. *Pandorina* is one of these 'plant-animals'. It is a cell-colony made up of 16 to 32 cells arranged in a gelatinous sphere. Each cell or zooid looks very much like *Chlamydomonas* and is able to manufacture its own food and propel itself by protoplasmic whips, or flagella. Coordination between the 16 is limited and *Pandorina* wobbles through the water in a rather aimless fashion, but it has got a 'head end' because one pole of the sphere always leads. When it is time to reproduce, the colony disintegrates completely. The zooids separate and each conjugates, like *Chlamydomonas*, with a zooid from another colony. The conjugated cells then divide by fission to form new colonies.

We can think of the cells of *Pandorina* as having learned to live together for most purposes but still having to reproduce as individuals. This situation has been overcome in *Pleodorina*, an organism very like *Pandorina*, but sometimes having 64 or 128 zooids. Some zooids at the 'head end' are smaller and more sensitive to light. They take no part in reproduction, so there is a definite separation into reproductive and non-reproductive parts of the body. Sex differentiation is also improving. There are male zooids which divide repeatedly to form mobile sperms and female zooids which do not divide. Each grows to form what is in effect an egg-cell or ovum.

There is, however, a big gap between these loose colonies of individually similar cells and even the lowliest metazoan animals proper, such as sponges or the many kinds of worms, but we can envisage a gradual reduction in the number of cells in a *Pleodorina*-like organism taking part in reproduction until special germ-cell tissues are recognizable, representing the ovaries and testes of higher animals. Meanwhile the non-reproductive cells are free to specialize in the formation of muscles, sense organs, and digestive and other tissues.

Is sex necessary? The advantage of asexual reproduction is its great simplicity. The animal just divides. Sexual reproduction requires that two individuals find each other, either to fuse, as in conjugation, or to let their gametes fuse and then to divide. Asexual reproduction can, therefore, proceed more rapidly and allow[illegible]mal populations to build up very [illegible] is discussed further in Chapter [illegible]tage of sexual reproduction lies in the mixing of genetic characters of both parents during fertilization, when the chromosomes in the nuclei, bearing the genes, are exchanged. The fusion of the sperm and ovum is the lynch-pin, the central event, of sexual reproduction. It is during this process that the individual characters of both parents are intermingled to produce a new and unique set of characters in the offspring, so that while there are family likenesses all members of a family are different. This is the starting point for natural selection, the mechanism of evolution. But, first, we must return to the binary fission of amoeba to understand the process of fertilization and cell division so fundamental to our subject.

When an amoeba splits in two, the daughter cells share the protoplasm and the two halves of the nucleus. The nucleus contains the chromosomes, microscopic rods of a complex material that contain the genetic blueprints for the entire animal. When an amoeba is about to divide, each chromosome forms an identical twin and as the nucleus splits, one half of each pair of these twins goes into each new nucleus. The two daughter cells, therefore, have an identical genetic constitution which is the same as that of their 'mother'. This process is called mitosis and takes place at every cell division in every animal except in the production of their sperms and ova for sexual reproduction.

In the formation of gametes or germ cells the cell divides twice but the chromosomes divide only once. This is called meiosis, a process markedly different from mitosis, and it results in the formation of four daughter cells, each with a single set of chromosomes. The daughter cells become the gametes and in fertilization, the nuclei of the two gametes fuse and their combined chromosomes arrange themselves in pairs. But the members of a pair are not identical. Each reflects the characters of the parent from which it comes, each parent having contributed half the chromosomes.

Sexual reproduction has therefore resulted in variation being introduced into an animal's offspring so that no two are alike. How the variation is controlled is the province of the science of genetics and heredity and need not concern us here except to note that the forces of evolution can now work on the varied offspring. Each animal is different from its fellows and some individuals will be better endowed than others. They have characters which give them a better chance of survival under adverse circumstances. Less fortunate individuals die while the survivors breed and pass on their useful characters to their offspring. This is natural selection, the process postulated as the main mechanism for evolution by Charles Darwin, in 1859 in his *The Origin of Species*, and by Alfred Russel Wallace in his *Contributions to the Theory of Natural Selection* published eleven years later. Without the variation caused by sexual reproduction natural selection could hardly operate.

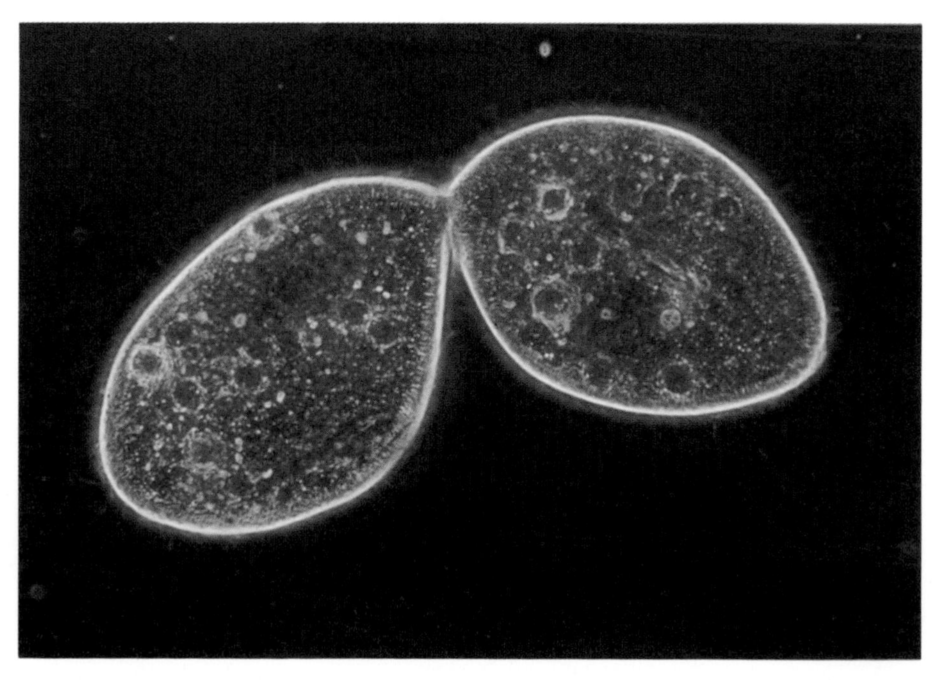

Paramoecium. Two cells in the late stages of division.

OF MATING AND MOTHERHOOD

The emergence of two sexes gave animal species great advantages in producing variation in the offspring so that as the conditions of the environment changed some would be more likely to survive than others. The big disadvantage was that for fertilization to take place, the gametes from each sex had to be brought together, the liberation of sperms and ova had to occur at the same place and time. This synchronization is effected in the higher animals by courtship. Even in the lower orders where it is hardly possible to talk of courtship in the usually accepted sense, there is of necessity a mechanism that brings the sexes together. Their reproductive methods are crude and consist merely of the sexes liberating gametes into the water in which they live, leaving the sperms to swim to the ova.

Sea urchins and mussels are typical practitioners of this method. If their gametes were released at random into the vastness of three hundred million cubic miles of ocean a sperm might have to travel a long way to meet an ovum. Since a sperm can swim no more than a few inches before it runs out of energy, fertilization would seem a little haphazard. Even under the best circumstances there must be an incredible wastage in eggs and sperms that never meet and fuse and so are lost, representing a loss in valuable food reserves that the parent animal could have used more profitably. Under the worst circumstances, that is, if there were no efficient way of bringing ova and sperm together, it is doubtful whether the species could survive. For example, each female mussel releases up to 25 million eggs: ample, one might think, to cover the seabed ten deep in mussels except that, apart from the eggs and sperms that never meet, from the moment of spawning every hand is against the mussel. Eggs, larvae and juveniles, as well as adults, are the food of a great variety of other animals and from the millions of eggs produced by a single mussel during its lifetime only one or two individuals will survive, themselves to breed.

Almost as soon as the first appearance of a true differentiation into sexes had occurred, some method must have been evolved for giving the maximum chances of sperms meeting the ova. This could have been accomplished in one of two ways, either by the development of sperms with a wide cruising range or by some means of ensuring that sperms and ova were liberated close together.

The first might have been accomplished by having the sperms of large size so that they carried the energy for relatively long journeys. There is, in fact, an ostracod, a simple form of crustacean, that produces giant sperms. This has puzzled scientists, although it may be that these giant sperms can keep swimming longer so giving a greater length of time than usual to seek out an ovum.

The simpler way of solving the problem, and the one usually encountered, is to use a spawning crisis. Many marine invertebrates, like mussels and Sea urchins, concentrate their spawning into a short space of time so that there is a certainty of both ova and sperms being in the water at the same time. The crisis is started by one Sea urchin, say, spawning. Then a wave of spawning spreads through all Sea urchins in the area. The stimulus is probably a chemical secreted with the sperm that diffuses through the water and into the bodies of neighbouring animals. Merely breaking open a Sea urchin and allowing its ripe gonads to fall into the sea is sufficient to start mass spawning.

A chemical signal from one animal that affects the behaviour of others is called a pheromone and pheromones are used throughout the animal world to bring the sexes together. Coordination of sexual activity by

LEFT: The Hatpin urchin is named after its long spines. All individuals along a length of coast ripen and release eggs and sperms into the sea together, to ensure the greatest chance of fertilization. Substances from one urchin diffuse through the water to trigger spawning among its neighbours.

FAR LEFT: Mussels spawn in spring with peaks occurring at new and full moons. The spawning of one individual initiates a general release of germ cells on such a scale that the water may turn cloudy.

pheromones is first seen in the protozoans. The pairing of *Paramoecium* and the clumping of *Chlamydomonas* mentioned in Chapter 1 is caused by a chemical secreted from cells in one batch or clone affecting those of the other clone.

The pheromone of *Chlamydomonas* appears to have only an attracting function but that of another organism *Volvox*, is more far-reaching. *Volvox* is a larger relative of *Pleodorina* and *Pandorina* and consists of thousands of *Chlamydomonas*-like cells embedded in a gelatinous sphere. A *Volvox* reproduces sexually through the fusion of freely swimming sperms with large, immobile ova, or asexually by budding, that is, forming new colonies by fission within the parent colony. It would be a waste for a colony to produce ova when there are no male colonies nearby but this problem is solved by a pheromone from a 'female' *Volvox* causing asexual colonies to convert the developing zooids in the buds into sperms, which can then fertilize the 'female'.

Massed spawning of worms. The close coordination of individual spawning, as in a spawning crisis, has been elaborated by certain marine worms, the bristleworms related to the earthworm but with a row of bristly paddles down each side of the body. In the tropical Pacific there is a bristleworm called the Palolo worm whose claim to fame is a finely synchronized massed spawning that is crammed into four nights' furious activity. The Palolo worm lives in burrows in coral reefs. As spawning time approaches, the rear half of its body undergoes a fundamental change. The internal organs degenerate and this half becomes little more than a bag of reproductive organs with two rows of paddles. When the gonads are ripe, the reproductive end breaks off from the body and swims to the surface, where ova and sperm are released.

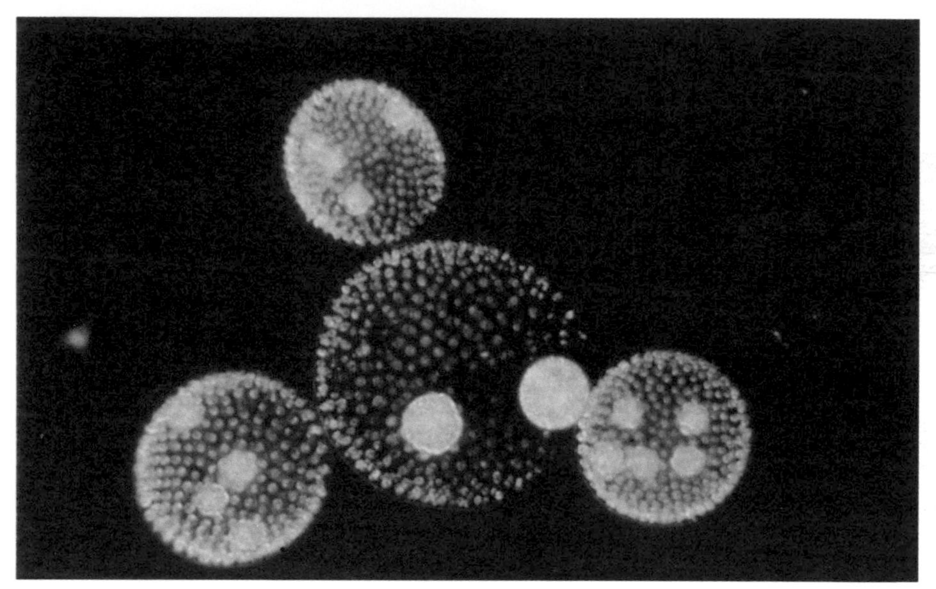

ABOVE: *Volvox*, a pin-head sized plant, showing the development of daughters. They grow inside its body and escape when it ruptures.

The shedding of the ova and sperm at the surface immediately helps to concentrate them in one place but they are also concentrated in time. Spawning takes place only at dawn on the night of the last quarter of the October and November moons and on the preceding dawn. So accurate is the spawning crisis that the South Sea Islanders know precisely when to launch their boats and find the sea writhing with worms and milky with masses of ova and sperm. The worms are scooped up and eaten raw or cooked.

The spawning of the Palolo worm, and of related worms that live in the West Indies and off Japan, is probably controlled by an accurate internal 'clock' which in turn is regulated by a combination of day and night changes and the monthly rhythm of tides. The Fijian islanders predict the spawning by watching for the flowering of certain plants, after which they look for the moon being on the horizon at dawn. Ten days later there is

Mbalolo lailai – the little Palolo, followed by *Mbalolo levu* – large Palolo.

The massive spawning crisis of the Palolo worm is a mechanism for concentrating the gametes of sedentary animals that live scattered through the coral reefs. It gives the ova a very good chance of being fertilized but can only work in favourable circumstances.

The ragworms, another kind of bristleworm, live on rocky or muddy shores eating small animals or filtering edible particles out of mud. They, too, form a special reproductive part as does the Palolo worm. At one time the reproductive stage of the ragworm was thought to be a completely different animal and was given the scientific name of *Heteronereis*. The name survives as a descriptive term. In many species the ragworms leave their burrows and hiding places at night and swarm at the surface in Palolo fashion. But in others the sexes come together more positively. Usually the males merely swim round the female *Heteronereis* in a primitive courtship dance but the American *Platynereis megalops* wraps itself around the female.

In the ragworms, therefore, we see the first steps towards a proper courtship. The males gather around the females and avoid the wastage of gametes by depositing sperms near the ova. There are in this behaviour signs of the essential pieces of information that must be passed between the sexes during courtship. One sex, at least, must signal its position and also its identity so that the second knows that it is courting an animal of the right species and sex. Finally, it must be established that both have ripe gonads and are ready to mate. At the simplest level, one signal is sufficient, as when the female makes a signal that is peculiar to her species and sex and shows that she is fertile.

Another bristleworm, the fireworm of the West Indies, demonstrates the use of synchronizing signals, which it combines with an internal clock mechanism like the Palolo worm. Christopher Columbus and his crew were the first Europeans to see the courtship of the fireworm although they did not realize the significance of what they saw. The fireworm spawns at the surface at set periods like its relative the Palolo worm. For about 20 minutes, starting 55 minutes after

Bristleworms, seashore-dwelling relatives of the earthworms, indulge in a primitive courtship. The sexes mass together for spawning.

RIGHT: The crayfish carries her eggs on her abdomen, where they are fertilized. After hatching, the baby crayfish spend some time clinging to their mother.

ovulate, that is, emit eggs from the ovary, for several days.

Enlarged claws are common among male crabs. Carrying the female during courtship is only one of their functions; one claw of the Fiddler crabs has become a gigantic semaphore signal. Fiddler crabs live on muddy beaches and estuaries around tropical coasts. Each crab has a burrow from which it emerges at low tide to feed and to which it scuttles whenever danger threatens. During its forays from the burrow, the Fiddler crab feeds on minute organic particles which it sifts from mud scooped up in its claws. Male fiddlers can only eat with one claw because the other is so huge that it dwarfs the rest of the animal. The claw is also brightly coloured and, from a distance, the flat mud surface of the shore seems to twinkle with flashes of colour as thousands of male fiddlers wave their claws. Each species semaphores its own particular recognition. Some wave like a departing traveller, others appear to beckon insistently and others again, constantly open and shut their claws like grotesque pairs of scissors. Then there are crabs that pose statuesquely with claw raised and trembling or dance from side to side. Whatever the form of the display, that of each species is so unchanging that it is easy to identify a species of Fiddler crab by the way it displays.

Charles Darwin watched Fiddler crab displays and came to the conclusion that they are performed to impress the female and that she is attracted to the male with the biggest and brightest claw. Other writers have repeated, and elaborated, this idea, and have described the courtship of Fiddler crabs in the most poetic and anthropomorphic language in which 'passionate' males display feverishly at the females and fight their rival suitors. Careful scientific observation has poured cold water on this romantic picture.

The main function of the claw-waving is nothing more than to signal ownership of a patch of mud to nearby males. Each male has his own precinct and fights other males, not for the attentions of the females, as was for so long supposed, but for ownership of the precinct. If an intruding male takes no notice of the resident, or is positively hostile, a fight breaks out in which the contestants grapple with their claws and push against each other. The fights are usually brief and the vanquished crab, normally the intruder, retreats.

By defending his small patch of mud, the Fiddler crab is free to court passing females without interference. When a female approaches, the male's displaying becomes more rapid and intense but she seems to take no notice at first. She appears 'coy' and this anthropomorphic term well conveys the slow

RIGHT: Many crustaceans carry their eggs with them; the Fairy shrimp carries hers in a large brood pouch. The male seizes the female with his antennae and transfers sperms to the brood pouch, where the eggs are fertilized.

LEFT: Gift-bearing plays a part in the courtship of the Banded coral shrimp. Here the male is approaching the female with a gift of food. If she is already feeding she will spurn the gift. Coral shrimps may mate for life.

Mating pairs of oceanic crustaceans. The courtship of crustaceans is varied, often bizarre. For many species the anatomy of the reproductive organs has been described but mating behaviour remains a mystery.

response of the female. She is, in fact making sure that she is being courted by a male of her own species because there may be a dozen kinds or species of Fiddler crabs living on the same beach.

As if claw-waving displays were not enough, Fiddler crabs also 'call' by rubbing their claws along a row of teeth on the edge of their shell. Crabs have no ears so it is to be presumed that they feel these calls as vibrations through the ground. The main function of the calls is to show to other individuals that the crab is in its burrow and is not to be disturbed.

The tropical Ghost crabs, so called because their grey shells blend so well with sand that when they stop running they seem to disappear, also call in this manner. Their burrows act as resonators to produce quite a loud chorus from a beachful of crabs. Again, deterring trespassers is the main function of their calling as far as is known but it would seem quite likely that females could respond to them as the female Fiddler crabs do to the visual displays of their males.

To sum up, in two major groups of animals, the marine bristleworms and the crustaceans, can be seen a trend of increasing sophistication in courtship and mating techniques. It must be stressed, however, that the descriptions given are merely examples of steps taken in this trend and do not indicate a particular line of evolution from one species to another. All that can be said is that these are the steps that were probably taken, by animals long extinct, in the search for improved reproductive methods. The first step in this trend is that sperms and ova are merely shed into the sea and left to their own devices. Later mechanisms of mating are improved and among the crustaceans, the boundary between external and internal fertilization is reached. In the former, the eggs are fertilized outside the body, as in crayfish, but crabs are on the way to developing internal fertilization. They place sperm in the female's body although it is later extruded with the ova and fertilization in fact takes place outside the body.

The tree-climbing Robber crabs of the Pacific and Indian Oceans are mainly land dwellers but they are still tied to the sea. They must return to it to breed. The only crustaceans to have overcome this problem and taken to a fully terrestrial life are the woodlice, whose rightful place is in the following chapter.

Initially, courtship ensures a synchronization so that gametes are shed in the same place at the same time. The use of signals such as pheromones and displays is an improvement that helps bring individuals of opposite sex together, thus increasing the chances of their gametes meeting and fusing. It also ensures that the two sexes are of the same species and in breeding condition. This is the level of behaviour reached by crayfish. Mating is a personal activity between two individuals, in contrast to the mass involvement of thousands, as in the Palolo worm. So courtship sets the stage for selection of a mating partner and it becomes possible for one animal to choose another as a mate and for several to compete to be chosen. This is what Darwin meant by sexual selection, which he defined as the struggle between individuals of one sex for the possession of an individual of the other sex. It becomes increasingly an important factor in the courtship as we move from the lower to the higher animals and is a potent force in the evolution of the species. Sexual selection consists of two parts; the struggle between males and the choice by a female of one of these males.

RIGHT: Fiddler crabs are the most spectacular of crustacean suitors. The males have an enlarged, coloured claw with which they signal ownership of a small territory and attract passing females. One beach may house a number of species of Fiddler crab each with its own different signal. This crab signals by raising its claw. As it does so it rises on tiptoe then flops down before repeating the gesture.

BELOW RIGHT: The male Soldier crab of mangrove swamps seizes the female and, while holding her down with some of his legs, quickly builds an igloo of sand over them with his remaining legs. Mating takes place inside, then the female is released.

BELOW: The West Indian Sand crab or Ghost crab signifies by stridulating that its burrow in the sand is occupied. The inner side of the large pincer bears a row of teeth which is drawn over the upper part of the limb to make a squeaking or croaking sound.

Abundant on both American and European coasts of the North Atlantic, the Edible periwinkle lives on the lower half of the shore. It practises internal fertilization but the eggs are shed into the sea and the larvae swim about for several days before settling.

INTERNAL FERTILIZATION

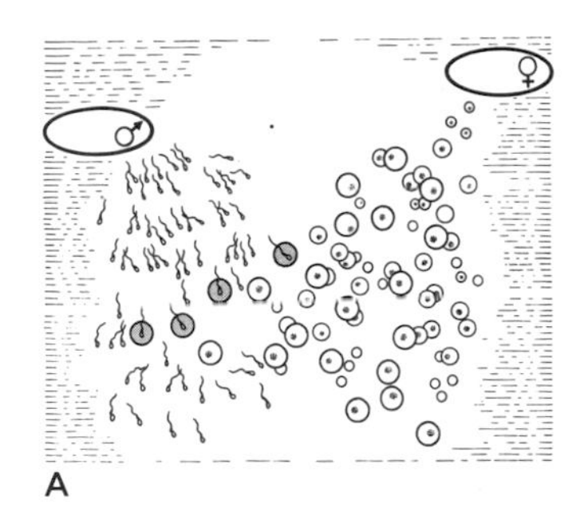

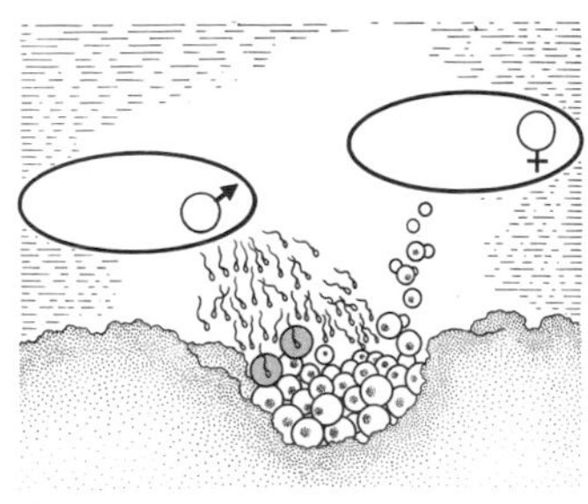

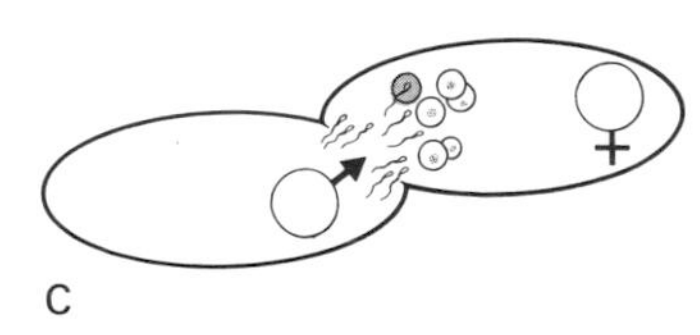

Animals such as many marine invertebrates and fish reproduce by external fertilization (A). Eggs and sperm are released into the water and fertilization takes place at random. Other animals, for example some fish and amphibians, increase the chances of their eggs becoming fertilized by depositing the eggs in a discreet spot such as a redd (B). The best chances of fertilization occurring are found in animals like mammals which employ internal fertilization (C) in which sperm are introduced directly into the female.

A brief survey of the development of sex in the animal kingdom shows a progression from the broadcast scattering of gametes into the sea to the strategic deposition of sperms as near as possible to the ripe eggs in the female's body. This must not be thought of as a steady development from the 'lower' forms of animal life to the 'higher' forms such as birds and mammals. Indeed, the concept of lower and higher animals is somewhat misleading although 'lower animals' is a term often used for those with a relatively simple anatomy and physiology. Nevertheless, internal fertilization, which is the fertilization of eggs before they are laid, is the most efficient mode of reproduction and, as we shall see, it has evolved in several widely different groups of animals. These animals can be said to be 'advanced' in this respect because internal fertilization confers several advantages. It usually, but by no means always, necessitates the male introducing sperm into the genital tract of the female by means of an intromittent organ – the penis, for instance, in mammals. But whatever the mechanism, wastage of both sperms and eggs is kept to a minimum and, by retaining the eggs in her body, the female can give them increased protection. Once this has been achieved the evolution of more developed systems of live birth, or vivipary, can begin. Or, if eggs are laid, they can be covered with a waterproof shell before leaving the female's body so that they can survive out of water.

The development of internal fertilization can be seen, therefore, as a major step in the history of evolution. It has opened the road to successful exploitation of life on land because both the gametes (sperms and ova) and offspring are protected from drying up. In turn, internal fertilization is made possible by the development of mechanisms for placing sperms into the female and of courtship behaviour for ensuring contact between the sexes at the appropriate time. Courtship has already appeared among animals practising external fertilization, as described in Chapter 2, but it has, in general, developed to a greater degree in animals practising internal fertilization. In this chapter we trace the development of infernal fertilization in two groups of animals as they make the transition from an aquatic to a terrestrial way of life and then review the different methods of attaining internal fertilization that are practised by the more advanced animals, so as to set the scene for the remaining chapters.

Woodlice and Periwinkles. The essential link between internal fertilization and living on land is seen in the phylum Arthropoda, the joint-legged animals. Among many crustaceans, such as crayfish and shrimps, fertilization is external, the sperms being stuck to the female's body near her genital aperture, so that they meet the eggs as they are extruded. However, in one group, the Isopoda, fertilization is internal. The Isopoda include the woodlice as well as the slaters of pond and seashore which can be best described as aquatic woodlice. Woodlice are the only wholly land-dwelling members of the huge crustacean class – there are land crabs but these must go to water to spawn – and internal fertilization is one of the secrets of their ability to live on land. They have a strange double copulation, each sex having two sets of reproductive organs. The male tests the breeding condition of the female with his antennae and, without further ado, draws to one side of her and, twisting the rear half of his body under her, slips the intromittent organ on one of his left legs into her right genital aperture. Copulation lasts for five minutes, then he moves to the other side and inseminates her left-hand reproductive tract with his right-hand organ.

The link between internal fertilization and terrestrial life is also demonstrated by the mating habits of periwinkles. These marine snails are often used as examples of an evolutionary stage through which what are now strictly terrestrial forms must once have passed. Although still tied to a watery life, they can live for some time out of water without drying up. On any rocky shore there is a change in the species of periwinkle from the lowest tide's edge to the top of the beach. Common and flat periwinkles live low down on the shore. They breathe by means of gills and cannot survive out of water for long. The rough periwinkle lives about halfway up the shore and the small periwinkle survives in the splash zone where it may not be wetted by salt spray for days on end if the sea is calm. Both have a lung chamber very much like that of the land snails to give them independence from water for breathing purposes. All four species reproduce by internal fertilization so that they can mate out of water or, if in water, minimize their gamete loss. Rough and flat periwinkles have also become less dependent on water in that the former's eggs are laid on seaweed and hatch as shelled young, omitting the free-swimming larval stage that is so common among marine invertebrates, and the latter's eggs complete their development within the mother's body. Paradoxically the small periwinkle still reproduces by larvae, and these must be shed into the sea to swim. In other ways however, it is the nearest to being a land-dweller.

Internal fertilization is by no means the prerogative of land-dwelling animals, as is demonstrated by the aquatic slaters, as well as by other aquatic animals, both freshwater and marine. Among the 'lower animals', some Sea anemones retain their eggs inside the body until after fertilization. They are fertilized by sperms swimming to them and entering the Sea anemone's 'stomach' through its mouth. The eggs develop into elongate, somewhat cylindrical, ciliated larvae, which later swim out through the mouth and eventually settle down and change into adult anemones. Hydra, the freshwater relative of the Sea anemones, also practises internal fertilization although, here, the eggs develop in ovaries on the outside of the body. Barnacles, being securely fastened to a rock or a ship's hull, cannot come together in copulation. We would therefore expect to find that, as with anemones, sperms are liberated into the sea to make their own way to the eggs. However, barnacles habitually live in tight-packed masses and fertilization is effected internally by means of a very long penis which can reach into the bodies of neighbouring barnacles.

Argonauts and Octopuses. If, to our eyes, the spawning of Sea anemones, oysters and sea urchins seems a cold, mechanical action of spewing out gametes into the sea with no contact between the sexes save for the exchange of some chemical stimulant, it is at

RIGHT: An unusual home for Acorn barnacles, which prefer to settle near their fellows. Inability to move and internal fertilization means that a barnacle can mate only with its closest neighbours.

Playing hard to get? One African woodlouse remains rolled up while another investigates it. The woodlice appear in huge numbers for a few days after the rains. They mate then disappear under cover, never to be seen again during daylight hours.

Normally solitary animals, squid gather for communal courtship and mating. The males court by displaying a sequence of colour changes.

least not so bizarre as the love life of squid, spiders, and their relatives, where the sperms are literally handed to the female in a packet. Squid belong to the group of molluscs known as cephalopods, which also includes the octopus and cuttlefish. Another cephalopod, distantly related to the octopus, is the argonaut or Paper nautilus. Argonauts live in warm seas and the females protect themselves and their offspring with a thin, papery shell secreted by glands at the base of two of the eight arms. In 1827, an Italian zoologist, Stefano delle Chiaje, caught an argonaut and found what he thought was a parasitic worm, later named *Hectocotylus*, which roughly means 'arm of a hundred suckers'. Only 26 years later was the '*Hectocotylus*' recognized as part of the minute male argonaut and the strange breeding habits of this mollusc set on record. Delle Chiaje's 'parasitic worm' is one of the eight arms of the male argonaut. The arm lies within a thin-walled sac on the male's body. A reservoir near its base is charged with sperms, after which the hectocotylus, as it is still called, bursts out of the sac to be revealed as an arm five times longer than the rest of the body, whiplike at the tip and bearing 50–100 suckers at the base. During mating, the hectocotylus is placed in the female argonaut's body. It snaps off and is left behind as the male swims away.

Only in two octopuses is there a detachable hectocotylus. In other cephalopods, the hectocotylus is used to place spermatophores in the female and is then withdrawn. Squid, such as the common American squid, court the

Mating in octopuses and other cephalopod molluscs is achieved by transferring a packet of sperms to the female by means of the hectocotylus, a specially modified arm.

female with a display of colour changes. Coloured spots appear on his arms and rosy flushes suffuse his whole body, in appearance rather like our own blushing. Eventually the pair copulate by lying underside to underside or head to head with arms entwined. Swarms of squid gather for communal courtship and egg-laying results in acres of the seabed being covered in strings of eggs. The sperms are bound in packets of mucus called spermatophores, each $\frac{3}{5}$ in (15 mm) long and surrounded by a horny casing. They are stored until used in a pouch in the mantle cavity, a deep cavity that contains the gills, ink sac and openings of the excretory and digestive systems and acts like a bellows squirting water through a funnel to propel the squid. During copulation, the hectocotylus, the fourth arm on the left side of the squid's body,

RIGHT: After laying her eggs in shallow water, the female octopus guards them until they hatch. Then she dies.

LEFT: Not such a frequent fate as is often supposed. The female Garden spider is eating the diminutive male but he usually takes care to impress upon her that he is a mate and not a meal.

BELOW: A more successful approach by a male Garden spider. By vibrating strands of the web in a 'signature tune' he identifies himself to the female and she allows him to advance. Before approaching the female's web, he will have spun a minute web on which he places a spermatophore. He picks this up with his pedipalps and transfers it to the female.

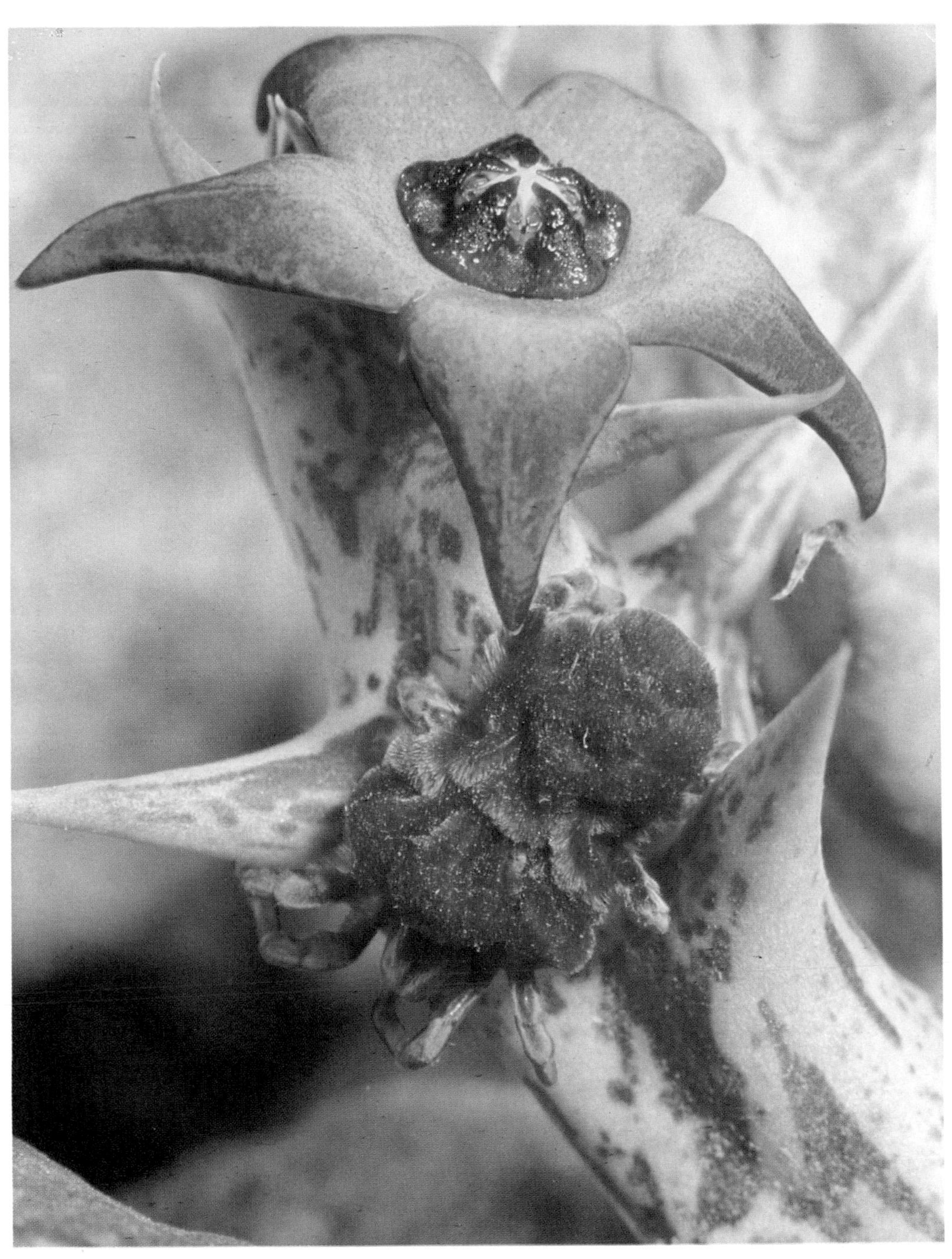

The African giant Red mites feed on plants and appear in vast numbers on the ground during the rains. Mating is a very casual affair in which the sexes do not even meet. The male leaves a spermatophore on the ground and the female picks it up later.

is drawn in front of the funnel and collects a bunch of spermatophores ejected from the mouth cavity. It is then thrust into the female's body. Sperms are released there by an elaborate mechanism. The cap of the spermatophore is torn off as it is pulled out of the pouch and sperm is forced out by contact with sea water. At the top of each sperm mass is a blob of cement which secures it to the wall of the female's mantle cavity until, over the next 24 hours, the sperms start to swim in search of eggs.

The mating of cuttlefish is a similar procedure, the male adopting a bright zebra-striped courtship dress. He swims above the female, stroking her with his arms. Copulation may not take place immediately but eventually he seizes her head and, while she throws back her arms, he transfers spermatophores with his hectocotylus.

Scorpions and Spiders. Male scorpions, relatives of the spiders in the class Arachnida, deposit a spermatophore on the ground during courtship and the females pick it up in their genital opening. The male grasps the female's palps – the limbs that look like the claws of a lobster – with his own and steps to and fro and side to side in a dance that may last several hours. In some species, the two scorpions grasp jaws or entwine their tails and, eventually, the male extrudes a stalked spermatophore which is cemented to the ground. He manoeuvres the female over it and the horny plates covering her genital aperture retract and the spermatophore is taken in. In one scorpion, the spermatophore has a trigger mechanism and is shot into the female. The eggs are fertilized within the reproductive tract and the baby scorpions are born alive.

Insemination by spermatophore is the general rule among the arachnids, the class that contains the spiders, mites, ticks and harvestmen as well as the scorpions. The harvestmen and a few of the mites possess an intromittent organ. In other mites there is no meeting of the sexes. The male deposits a spermatophore on the off-chance that it will be eventually picked up. Male spiders and ticks, on the other hand, actually place the spermatophore in the female's body. As spiders are carnivorous, the male has to ensure that the female recognizes him as a potential mate rather than as potential prey. Males are often considerably smaller, so they run a definite risk of being eaten but the idea of the female spider being 'more deadly than the male' and invariably eating her suitor is not borne out by observation. The whole aim of spider courtship is to switch off the female's hunting behaviour and make her receptive to mating. So, providing he behaves properly, the male runs little risk.

For our purpose, the spiders can be divided into three groups based on breeding behaviour. These are the orb-web builders, the short-sighted and the long-sighted spiders. In each group, there is an elaborate ritual of courtship to woo the female. And in each group the ritual is geared to the dominant senses. But before courtship begins, the male spider prepares for mating. First he spins a minute web, only a few millimetres across, on which he discharges the spermatophore. He picks it up in one or both of his pedipalps, the limbs which spring from the body in front of the first pair of legs, and is ready to seek a female.

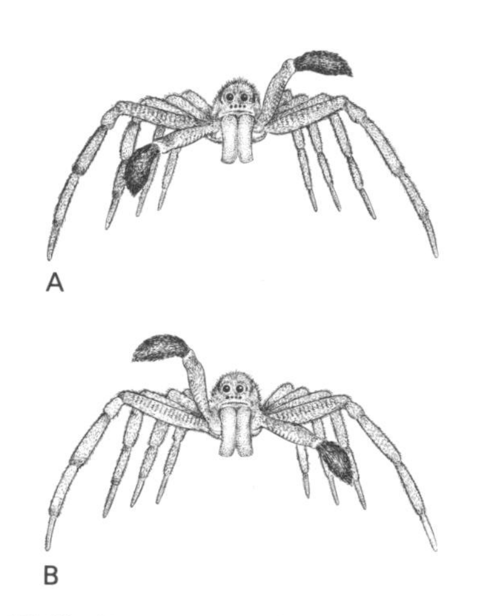

Wolf spiders have good eyesight and the males employ visual signals to attract the attention of the female. The pedipalps, limb-like appendages at the front of the body, are large and are used to 'semaphore' signals. Two positions of the pedipalps are shown here (A and B).

Scorpion courtship involves a lengthy dance that ends with the male dropping a spermatophore and manoeuvring the female over it. She picks it up and the young scorpions are later born alive. They spend the first part of their lives on their mother's back.

Safe in their silken cocoon, the eggs of the Wolf spider are carried by their mother. If the cocoon is wrenched from her, she searches for it frantically.

Quite how a male spider tracks down the female is not known but the male orb-web building spider eventually arrives at her web. Standing at the edge, he plucks the strands in a 'tune', a set sequence of vibrations, that identifies him to the female. Presumably her hunting behaviour is thereby switched off and she is prepared for mating. Gradually he draws near and, after caressing her with his legs, straddles her and transfers the spermatophore. Some spiders wait until the female is busy devouring an insect before advancing, while others boldly grab her jaws in their own, so thwarting any hostile action.

A large number of spiders have adopted an active way of life by hunting their prey rather than by setting a snare in the form of a web. Among this number are the huge, repulsive Bird-eating spiders, the Jumping spiders, the Wolf spiders that carry large cocoons of eggs on their bodies and the hairy House spiders that so frequently get trapped in empty baths and basins. These spiders hunt by sight, each one having several pairs of eyes. They also employ vision in courtship.

The long-sighted spiders use visual displays to court the female while still at a safe distance. They use the visual equivalent of the web-builders' 'morse code' of vibrations; each species 'semaphores' his identity by waving his pedipalps or a pair of legs, or both. Very often these limbs are enlarged or coloured to make them more conspicuous. Some of the Jumping spiders dance to and fro on tiptoe, then, as the female becomes receptive, the male inches forward until he can complete his wooing by caressing her with his legs. *Pisaura listeri* uses a gift in his courtship. He catches an insect, wraps it in silk and presents it to the female. Some of the Wolf spiders accompany their courtship dance with music. Strictly percussion, the accompaniment is made by drumming the ground with the pedipalps and legs and the sound can be detected by the human ear. *Lycosa rapida*, an American Wolf spider, waves its pedipalps alternately in a circle then taps and drums with both pedipalps and the first pair of legs. If the female is receptive, she signifies her acceptance by raising and lowering her forelegs.

For the other group of hunting spiders, the short-sighted spiders, vision alone is not sufficient to ensure the male's safety. Their eyes are adapted for recording movement, so the stationary female will spot the searching male first and treat him as prey. Consequently he must locate her position by smell and start the all-important identifying signals before she attacks. He quickly grapples with his intended, perhaps seizing her by the leg, while he uses his own limbs to caress and subdue her. Some species indulge in unusual stratagems. In the fierce *Drassodes* spiders, the

ABOVE: Crab spiders are short-sighted. They lie in wait for their prey to fall into their clutches and the miniscule male Crab spider must make sure he does not follow suit. He signals to his potential mate and may even grapple with her so that she cannot overcome him.

LEFT: The tarantula's means of mating without death ensuing is to hold his partner's fangs with special spurs on his front legs.

male seeks an immature female and imprisons her until her final moult, when he will mate and leave her before she has reached her full strength. The males of *Xysticus* spiders tie their mates to the ground with silken threads.

The principal characters in this chapter, the woodlice, periwinkles, cephalopod molluscs, scorpions and spiders, are unrelated animals that have independently developed means of internal fertilization whereby sperms are transferred from male to female without danger of loss. The same is true of the land snails and slugs whose reciprocal courtship and mating will be described in the next chapter. The myriapods, the group of many-legged animals, comprising centipedes and millipedes, are yet another group of land-living animals with internal fertilization. The carnivorous centipedes, characterized by one pair of legs per body segment, deposit spermatophores in the manner of scorpions, while the plant-eating millipedes, with a basic arrangement of two pairs of legs on each segment, have intromittent organs for transferring sperm.

The many-legged millipedes couple by means of an intromittent organ, a more intimate mating than that of the fewer-legged centipedes which deposit a spermatophore on the ground.

Although the mating habits of all these animals appear quaint and perhaps over-elaborate to our eyes, they have one important feature in common. They are all 'higher' animals in terms of body organization. They are actively mobile, with the exception of the periwinkles and land snails which are hardly fast movers, and they have well-developed nervous systems and sense organs. Each group, in its own particular way, represents a successful form of life. These characteristics go hand in hand. Internal fertilization gives an animal a certain independence and control of its environment, as do mobility and well-developed sense organs. Mobility and sensory capacity also allow the elaboration of the courtship behaviour that plays an important role in bringing the sexes together.

Insects and Vertebrates. Everything said about the other animal groups applies equally, or more so, to these two groups, arguably the most successful of animal forms in terms of the numbers of ways of life that they have exploited. It is among the insects and vertebrates that we find the greatest development of mobility and sensory capacity, as well as the greatest diversity and complexity of courtship behaviour.

In both groups we find copulation with insemination through an intromittent organ to be the general rule. Exceptions among the insects are the primitive wingless bristletails and springtails, in which the male deposits a stalked spermatophore for the female to pick up in the manner of a scorpion. The male sex organ of insects lies near the tip of the abdomen, on the underside between the ninth

Male *Sepsis* flies court among low-growing vegetation, then carry their mates to a suitable cowpat where mating and egg-laying takes place.

and tenth segments. At rest it is retracted into the body like a glove finger which has been pulled inside out. When mating takes place, it is everted to form a rigid tube. The proper name for this organ is the aedeagus but it is often referred to as the penis. Sperms are inserted into the vagina, the end portion of the oviduct, either in a spermatophore packet or freely swimming in a seminal fluid.

In the vertebrates, external fertilization is practised in the two aquatic classes, the fishes and the amphibians. But there are exceptions to this. The cartilaginous fishes – the sharks, rays and their relatives – have intromittent organs composed of modified parts of the pelvic fins. Called claspers, it was once thought that they were used for holding the female during copulation but now it is known they form an intromittent organ when pressed together and inserted into the female's cloaca, the duct that forms the common terminus of the alimentary, excretory and reproductive system.

Guppies and related toothcarps of tropical fresh waters have a gonopodium, a long tube formed from part of the anal fin. The genital opening lies just in front of the anal fin; the pelvic fins are used to wrap over the aperture and so guide the sperms into the gonopodium.

A unique elaboration of the gonopodium is found in the Four-eyed fish of Central America, so named for the division of each eye into two halves, the upper for vision in air, the lower for vision in water. The gonopodium can be moved either to the left or to the right, but not both ways as can the gonopodia of other toothcarps. The females also have a scale on one or other side of the genital opening, so 'left-handed' males can mate only with 'right-handed' females and vice versa. There are equal numbers of left-handed and right-handed individuals in the wild but the purpose of this 'handedness' of reproductive organs is a mystery.

External fertilization has tied the amphibians to water. They breathe air and their skin resists desiccation but they must return to water to breed. There are a few exceptions. Some of the newts and salamanders have developed internal fertilization and land-breeding occurs in some species. Only one species of amphibian, the Tailed frog, has an intromittent organ and it still breeds in water. The Tailed frog is named after the tail-like appendage of the male that is, in fact, an enlarged exterior cloaca that serves to introduce sperms into the female's body. This unique organ may be an adaptation for life in the mountain torrents of northwest America in that it prevents sperms being swept away.

Reptiles and Birds. Unlike the amphibians, the reptiles have severed all links with their watery past. Fertilization is effected by a penis, the sperms are transferred to the female in their own fluid medium and, after fertilization, the eggs are retained so that they can be covered with an impervious leathery skin to prevent their drying up. The penis of reptiles takes a variety of forms. The tortoises and turtles have a penis that lies on the floor of the cloaca. The lizards and snakes have paired copulatory organs called hemipenes that retract into the tail when not in use. At one time it was thought that the two hemipenes were pressed together for intromission, in the manner of the claspers of sharks, but in fact they are used singly. When erected, by muscular contraction and blood pressure, the hemipenes stand out laterally. Which one is used depends on which side of the female the male approaches.

The birds are rather a puzzle. Most lack a penis and copulate by pressing their cloacae together. Yet the reptiles, their ancestors, possess a penis, except in the case of the

RIGHT: Most fishes lack an intromittent organ but male mosquitofish, the lower two in this picture, bear intromittent organs, or gonopodia. These are formed by the anal fins and are used for transferring sperm to the reproductive tract of the female.

Copulation in birds is usually achieved by the male and female pressing their cloacas together. Only swans and a few others possess a penis.

tuatara of New Zealand. The situation is complicated because the penes differ in structure in the several groups that possess them. Penes are present in four groups of birds, notably the ducks, geese and swans and the ratites – the ostrich, rheas, emu, kiwis and cassowaries. In the act of copulation of most birds, the feathers surrounding the cloaca are turned back and, by turning the tail down and sideways, the mounted male brings his cloaca in contact with that of the female. The lips of the cloacas are everted so that the openings of the vasa deferentia, the tubes which carry the sperms from the testes in the male's cloaca, are brought into direct contact with the mouth of the oviduct in the female's cloaca.

Mammals. In mammals, the penis is a development of the reptilian organ. It consists of spongy tissue which increases in size and rigidity through engorgement with blood and is sometimes supported by a bone, the baculum. The penis of some mammals is always visible but in seals, for instance, it retracts completely within the body. The sperms are carried from each testis in the vas deferens which links with the ureters running from the kidneys to form a common duct, the urethra.

Among the so-called 'lower' animals, mating is an automatic, matter-of-fact process.

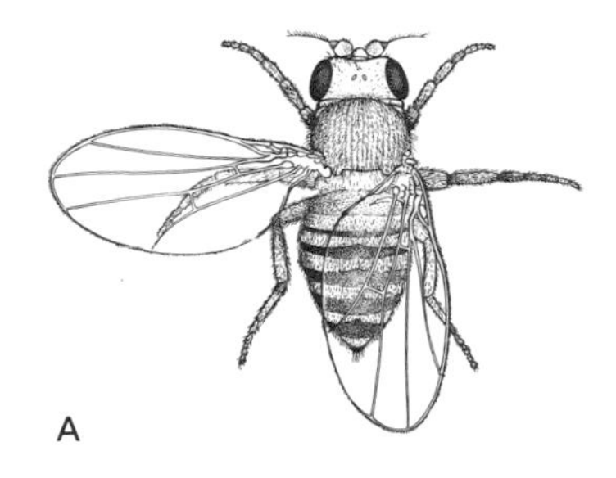

To attract its mate, the male fruit fly extends one or both wings (A) and vibrates them to produce a 'song' composed of a pulsating buzz. The female is able to distinguish between the pulse rates of different species, a difficult task as is shown by the oscilloscope records in (B). The pulse rate of one species, *Drosophila persimilis* (1), is three times as fast as that of *Drosphila pseudoobscura* (2). The slower pulse of *Drosophila pseudoobscura* occurs at a rate of one every fifth of a second; the scale at the bottom of the diagram indicates divisions of hundredths of a second.

If two animals meet and recognize that they are of the same species, opposite sex and in breeding condition, they mate and separate. It is as simple as that. The male woodlouse sounds out the female with his antennae and, if she fulfils the necessary requirements for woodlouse mating, he moves alongside and inseminates her. But as we move to more 'advanced' forms of life, courtship behaviour begins to lose its stereotyped, unvarying appearance. It would be taking things too far to suggest that the protagonists in a courtship encounter show any freewill but at least they do not always react in the same way. In the courtship of cuttlefish, the zebra-stripe display does more than demonstrate the sex of the animal. It serves to form a bond, albeit temporary, between the two and does not necessarily lead to copulation. And in the squid there are signs of growing excitement, with quick movements of the arms and sharp swimming movements. Presumably these are signals to show the partner that the squid is getting ready for mating. It is also significant that a male may persistently chase one female even when others prepared for mating are present. There is an element of choice in his behaviour which is lacking in the 'lower' levels of life. His behaviour is not just switched on or off by well defined signals; his brain is making its own simple decisions. This is an important point to bear in mind during the following discussions of courtship in insects and the vertebrates. It emphasizes that in mating behaviour we are not concerned merely with the behaviour shown by an individual animal but with the interaction of two individuals. Their behaviour makes up a single entity and, as animal behaviour comes under increasingly close scrutiny, we find that we must not even consider isolated pairs of animals but look broadly at groups or societies.

In the last three chapters in particular, we shall see that, while mating directly concerns two animals, others are involved in the events that lead up to courtship, especially in the choice of which individual is to be courted. The failure of some female squid to attract the male is as important as his chasing the favoured one. What was it that made him prefer one female to the rest? This is not a question that can be answered easily even for animals that have been better studied than squid, or even for the choosing of mates by human beings. Neither do the observations on the persistent male squid give an answer to the corollary: How many of the female squid would have accepted the male if he had tried to court them? A superficial view of courtship suggests that, as the male takes the initiative, he makes the decision as to the suitability of the pairing. This is not the case, as is well illustrated by the courtship of Fruit flies.

Fruit Flies. Also known as Vinegar flies, Pomace flies or Garbage flies, these are small insects whose larvae feed on fruit and cause considerable damage to fruit crops in warm parts of the world. The adults are also attracted to stored fruit and alcoholic drinks, thereby making a nuisance of themselves. Under their scientific name of *Drosophila*, Fruit flies are better known as the subject of experiments in genetics. There are about 2,000 species, some having a global distribution yet, despite considerable overlap in ranges and habitats, the species do not interbreed. They are kept separate by the specific nature of the courtship ritual.

A male Fruit fly has a very liberal approach to courtship. He will court not only any Fruit fly of any species but also any moving object of small size. His technique is to approach within two millimetres of the object of his interest and investigate it by means of smell and sound. He extends one or both wings and vibrates them, producing a buzz so faint that it can be heard only by placing a fruit fly on the diaphragm of a sensitive microphone. The buzz is made up of short pulses of sound of very definite duration and frequency. Each species of Fruit fly has its own buzz. The female senses the buzzing through her antennae and if it is the 'signature tune' of her own species, she allows the male to lick her genitalia and copulate. If the buzz is foreign, or if she is immature or has already mated, she dissuades her suitor by buzzing back and kicking him with her hindlegs. In this manner, the integrity of the species is maintained by the female's recognition of the male. The mechanism of recognition is so precise that the length of time between the pulses in the 'signature tunes' is sufficient to make an effective distinction between species. *Drosophila pseudoobscura* emits pulses at intervals three times as long as those of *D. persimilis* and the two species are prevented from interbreeding even though they are so alike that trained scientists, familiar with both species, have great difficulty in telling them apart. That they do not interbreed is a result solely of discrimination by the females. To rephrase the proverb: Man proposes, woman disposes. This is as true for squid, insects and spiders as for human beings.

The Fruit fly *Drosophila* attracts a mate by extending one or both wings and vibrating them to produce a 'song' composed of a pulsating buzz. The female is able to distinguish between the pulse rates of different species and so select the correct mate.

VARIATIONS ON THE THEME

Perhaps it would be reasonable to suggest that most people tend to take the existence of two sexes too much for granted. The attraction and meeting of male and female, the insemination of females by males and the fusion of sperms and ova appear to be universal among the 'higher' animals at least. Once a high level of bodily organization has been reached, animals seem to give up the cruder methods of reproduction by budding and fission and settle down to what we think of as a normal sex life. Yet there are exceptions. Some animals are bisexual, functioning as male and female simultaneously or alternately. Others avoid sex by omitting fertilization and reproducing by virgin birth.

Solitary sex and sex changes. We have seen in Chapter 3 that internal fertilization, the most efficient method of sexual reproduction, depends on a measure of mobility and a good sensory capacity. Some animals, for various reasons, lack mobility at least. In them the meeting of the sexes cannot be easily engineered and one solution to this problem is for each individual to carry the reproductive organs of both sexes. This is called hermaphroditism after Hermaphroditos, the son of Hermes the Greek messenger of the gods, and Aphrodite, the Greek goddess of love. Hermaphroditos was beloved of a nymph in the fountain of Salmacis. Her love being unrequited she prayed to the gods, as she clung to him, that they should never be parted and the gods made them one.

Hermaphroditism in its less romantic but more realistic form may be realized in one of two ways. The animal may fertilize itself so that it has no need to meet another animal for mating; or it may court and mate with another animal. In the latter case, each animal may function as both male and female. They exchange sperms, so simultaneously fertilizing each other's ova, as in slugs, snails and earthworms. Other hermaphrodites are first one sex, then change to the other, so at some stage in its life the animal is male and fertilizes females and at another period it is female and is itself fertilized. Self-fertilization is rare but occurs among animals that are completely incapable of movement and cannot seek a mate.

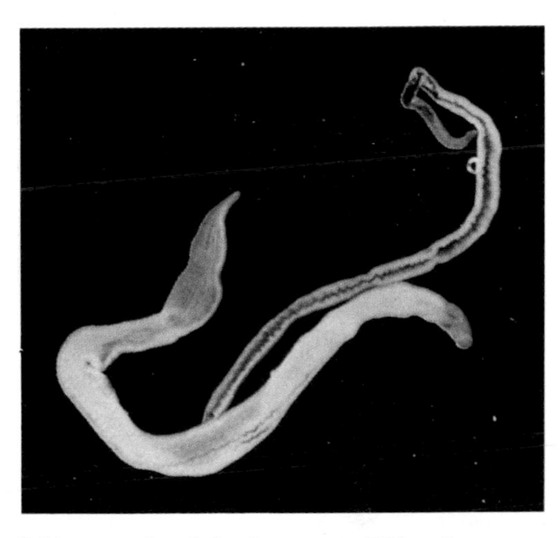

Schistosoma is a fluke that causes Bilharzia. Mature adults live in the veins of the human host and once a male has found a female, he clasps her in his genital groove. Huge numbers of eggs are laid into the blood stream.

However convenient it may be in other ways, self-fertilization defeats the object of sex. If an animal fertilizes itself there is none of the mixing of genetic characters that leads to variation in the next generation, so the species is denied the chance of rapid adaptation to a changing environment.

Among the few animals that practise self-fertilization is the tapeworm. Indeed, this is essential since often there is only one worm in the host animal. Tapeworms live in the intestines of many vertebrate animals. As the name suggests, the body is flattened and ribbon-shaped. There is a minute head end or

Imported from North America to Europe, where it has become a pest through competing for space in oyster beds, the Slipper limpet changes sex as it grows older. All Slipper limpets start life as males then become females. As they settle one on top of the other, the lower older animals are female and the upper younger animals are male. Those in the middle are of intermediate sex.

scolex, equipped with hooks, which fasten to the intestine wall and, growing from it, a series of segments, called proglottids are continually budding off the scolex and adding to the ribbon. Each proglottis contains both male and female organs but cross fertilization is not feasible because this would require a long, unwieldy intromittent organ, and if the sperms were merely to be shed into the intestine they would be quickly destroyed by its digestive enzymes. As it is, there is the very convenient arrangement that the female and male ducts open into a common cup-shaped outlet and self-fertilization is accomplished by sperms swimming the short distance from the male aperture to the female aperture. The fertilized eggs develop inside the proglottis, which degenerates into an egg-filled bag and drops off the tapeworm. Alternatively they are coated in a resistant covering and are shed individually into the intestine. In either case, the eggs pass out of the host's body before hatching.

Apart from tapeworms and a few other animals, hermaphrodite animals take particular precautions to avoid self-fertilization This is usually achieved by an individual functioning first as one sex then as the other. Either the testes ripen before the ovaries or the sex organs themselves change sex, producing first one kind of gamete, then the other. There is a small Sea slater, *Hemioniscus balani*, for instance, which parasitizes Acorn barnacles, feeding on their tissues. When young, it is male and can move about. Later it changes to female and becomes sedentary. While male, it travels from one barnacle to another, feeding on them and fertilizing the sedentary females living in them. When it finds a barnacle that is free of slaters it settles, turning into a female to await a visit from another male.

The Slipper limpet, a native of North America, expressively named *Crepidula fornicata*, was accidentally brought across the Atlantic in a consignment of oysters in the 1880s and it has become a pest in oyster beds. Slipper limpets are not true limpets but relatives of periwinkles shaped like limpets. They live, one on top of another, in piles, usually referred to as chains. All Slipper limpets start as males but when one settles on a rock by itself it quickly becomes a female. Eventually a second settles on the shell of the first and stays a male, fertilizing the female under it, using an intromittent organ. As each fresh Slipper limpet joins the chain, the male it settles on begins to change sex, ending as a female and being fertilized by the newcomers, so there are always females below, one or more males on top and those in between in process of changing from male to female.

Even Common limpets sometimes change sex although the reason is not yet known.

Slipper limpets have, however, good reason for doing so because they practise internal fertilization yet are incapable of movement except when immature. The ability to change sex ensures that a chain always includes at least one male perched on the females to effect fertilization.

Reciprocal mating. Hermaphroditism is common among the creepers and crawlers. Meetings between two animals may be uncommon but if each is bisexual, the chances of fertilization are more than doubled, as all contacts can lead to a mating. Instead of there being just a chance of the pair being of opposing sex and only one being fertilized, both are fertilized and two sets of offspring result from the single encounter. Some of the most familiar garden animals are bisexual crawlers.

On a warm, moist evening a visit to a lawn will reveal the mating of earthworms. Taking care not to disturb them with heavy footfalls or too bright a light, it is possible to approach close enough to see that each earthworm is about half way out of its burrow, with its rear half anchored in the burrow ready to spring back in a split second. Each is joined by its head end to another, similarly placed worm. How the two find each other to make contact is not known. An earthworm's sense organs are not well developed; presumably it senses traces of slime left by the movements of the other worm or perhaps they feel about in a circle around the burrow until they make contact. However, once contact is made, the two lie head-to-tail while bound together with mucus. The mucus is secreted from the clitellum or saddle, the broad band which is sometimes mistakenly thought to be the result of injury and regeneration. While the worms are held together, sperms issuing from the genital opening of each worm pass along a groove in the skin to the other worm where they enter through a pair of tiny pores and are stored in two small bladders, the receptacula seminis.

Mating of earthworms takes three or four hours, a surprisingly long time, but once finished the earthworms separate and do not mate again that year. They return to their burrows, inseminated but not yet fertilized. The sperms may not be ripe which is another advantage to a slow-moving animal. When a pair makes contact neither has to have ripe gametes so the maximum advantage is taken of the rare meeting. When the ova ripen in due course the clitellum secretes a girdle-like cocoon which slips forward, receiving ova and sperm as it passes over the openings to the ducts from the ovaries and from the receptacula seminis. Finally, the cocoon is shrugged off the head end of the worm and is left behind in the earth. The eggs are fertilized inside the cocoon and the larvae hatching from them later escape from the cocoon.

The mating of earthworms is a head-to-tail, reciprocal event. Earthworms are hermaphrodite; each individual is both male and female so that the pair fertilize each other and both lay eggs.

Soon after they have come out of hibernation, Roman or Edible snails mate. As they hibernate in clusters they are conveniently near each other and can mate before dispersing. Each snail is both male and female; they are true hermaphrodites that fertilize each other's eggs. Some of the Pond snails can fertilize their own eggs, and may do so even when they have mated with another snail.

The mating of snails has an unexpected touch. At the height of courtship when the snails have been pressing against each other for some time, each stabs the other with a 'love dart'. The dart may become completely buried in the snail's flesh but it does not seem to cause any great harm.

Slightly more active but still in the crawling class, the snails and slugs are hermaphrodites which embark on some strange courtship rituals. Molluscs had separate sexes during the early stages of their evolution and they discharged their gametes into the water at random. Some retain this form of reproduction while others have developed internal fertilization either between separate sexes or as hermaphrodites, as in the periwinkles.

The courtship of snails is at once simple and complicated. It is simple in that once a snail has met another the two come into contact without preliminaries, but once together, copulation may take some time and may require physical wounding of one partner.

Snails are hermaphrodite in the truest sense; they are male and female at the same time and mutually fertilize each other. When they meet, they rear up and press the flat undersides of their bodies together and caress each other with their tentacles. Several hours may pass in this manner, the snails swaying from one side to the other and sometimes forcing each other back into the shell. There is a copious secretion of slime which could help glue the two together or perhaps stimulate the partners chemically. Eventually the pair come to the height of stimulation by stabbing each other with chalky 'love darts', about $\frac{1}{3}$ inch long and contained in a sheath until ejected. A love dart may just graze the skin of the snail it is aimed at and fall to the ground or it may penetrate the skin and work its way deep into the body. This Cupid-like procedure is a most odd method of sexual arousal but it has in addition an identificatory function. For instance, Brown-lipped and White-lipped snails are extremely similar in appearance and share the same habitats, but interbreeding does not happen. The White-lipped snail has a small, curved 'love dart' that is ineffective when discharged into a Brown-lipped snail. Moreover, a White-lipped snail may be dissuaded from completing courtship with the Brown-lipped snail by the vigorous movements and more severe stimulation of the latter's larger, straight, dart.

The final act of courtship in snails is the mutual insertion of the penis of each into the female opening of the other. Sperms are exchanged and the two snails retire. Mating takes place after emergence from communal winter quarters, when the snails are still near each other, an important consideration for a slow-moving animal, after which they scatter over the countryside to lay their eggs. The sperms from the mating survive for several months and egg-laying continues until late summer.

The courtship of slugs involves the secretion of copious floods of slime, seen here as a pale, shining ball. The slugs eat each other's slime and Great grey slugs go to the extraordinary length of dangling together from a rope of slime before mating.

If the spearing of one's partner with a 'love dart' seems a perverted kind of courtship, the habits of some slugs are even stranger. Slugs are virtually shell-less snails although some species have a small shell which may be hidden inside the body. In courtship, they circle each other while eating each other's slime trail but the four-inch European Great grey slug and the large Black slug of America bring their courtship to a climax by a feat of acrobatics. The pair spend an hour or more circling head to tail on a tree trunk or wall and work up a copious mass of slime. Suddenly, they twine round each other and throw themselves into the air. They are now dangling at the end of a rope of slime and they spin to and fro while each unfolds a two-inch-long penis. These they wrap around each other and sperms are exchanged. When the performance is over, the slugs either drop to the ground or climb back up the slime trail.

Among Land snails cross-fertilization is the rule but slugs sometimes fertilize themselves, as do certain Pond snails. Copulation in Pond snails is rather different from that of Land snails. There is no 'love dart' and one snail acts as male while the other is female. The male chases the female, climbs onto her shell, stimulates her and then inseminates her. Pond snails are also hermaphrodite so the gender of the two is relative and it sometimes happens that during the courtship the 'female' meets a third snail and climbs on its shell. 'She' now behaves like a male towards the third snail while still being a female to the first. Theoretically, a chain could be built up, even to the extent of forming a ring of snails. Copulation is, however, not essential because

Pond snails kept in isolation breed successfully by fertilizing their own eggs. The advantage is that a Pond snail transported, perhaps by a bird, to a pond otherwise empty of snails, can soon found a flourishing population.

Dwindling males. In the hermaphroditic condition of earthworms and snails sex, or gender, as we know it, tends to lose its meaning. Maleness with its emphasis on action – finding and courting the females and repelling other males – has lost its emphasis. The trend has been taken even further in some animal species where males have become very rare or missing altogether.

We are so familiar with the male being the larger, more powerful sex that it seems strange to find males that are mere dwarfs and completely overshadowed by the females in size. The female must always be of a good size because she has to have food reserves for the manufacture of eggs but the male has only to produce short-lived sperms which require little energy. Thus, unless there is competition among males which leads to the evolution of 'giant' males as in baboons and sealions, the male can become little more than a mobile reproductive organ. He needs only enough food to keep him alive while he manufactures sperms and finds a female to inseminate.This is the situation with some spiders, in which there is great disparity in size between male and female, but it has been taken to extremes in some marine animals where the diminutive male has become a parasite on the female and is no more than a bag containing reproductive organs.

A strange-looking worm, *Bonellia*, living in rock crevices in warm shallow seas, belongs to the echiuroids, worms with no common name which are probably distant relatives of the earthworms and bristleworms. *Bonellia* has been rightly described as being the size and shape of a cocktail sausage, but vivid green in colour. Unlike the cocktail sausage it has a proboscis, forked at the tip and up to two feet long, with which it picks up its food. This worm seems always to be female because the male is a minute creature living inside the female. Male larvae settle on the female's proboscis and are carried back to her mouth as if they were food but, instead of being digested, they change into adults. They crawl out of the mouth and wander over the female's body until they find the opening to her reproductive duct. There they gather, perhaps 10 or more together, living in the female's brood pouch where they fertilize the eggs as they pass down from the ovary.

Stranger still, perhaps, are the Deep-sea anglerfishes living in the pitch-dark depths of the sea, where there is only the sparsest life. Because there is so little food they have to take every opportunity of eating. For this they have huge mouths and long, needle-like teeth that instantly trap any animal that comes within range. There is a similar problem in the matter of breeding. The fishes are so thinly scattered through the three hundred million cubic miles of ocean that every meeting of male and female must be exploited.

When Deep-sea anglers were first brought up in trawls they puzzled scientists because all were females. Eventually someone noticed that each one bore fleshy growths on the body. These turned out to be the males, completely parasitic on the females. Young male deep-sea anglerfishes are entire fishes, like the females but smaller. They roam the depths and when they meet a female, they make fast by biting into her flesh with sharp teeth and fuse with the female. The two blood supplies link up so that the male draws nutrients and oxygen from the female and his body degenerates until all that remains is a pair of testes and the female becomes, in effect, a hermaphrodite. One female may

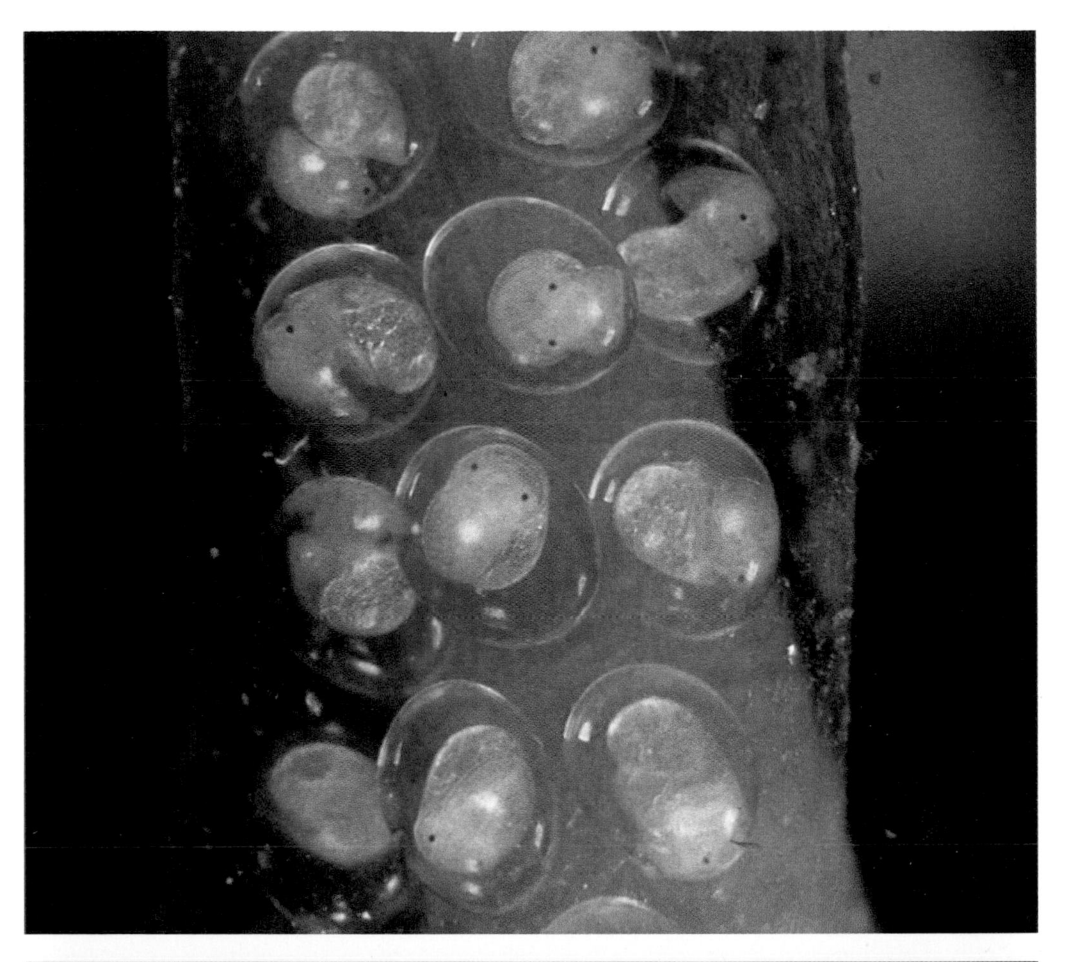

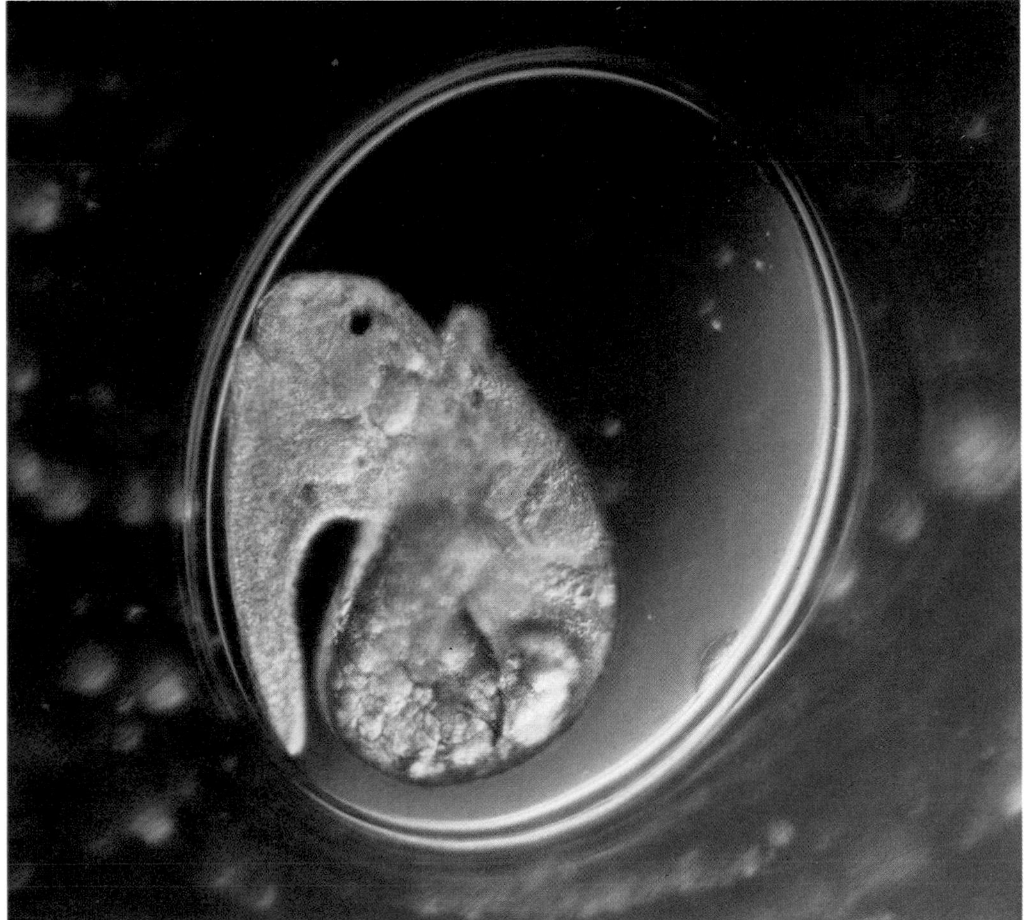

The eggs of Pond snails are laid in ribbons of jelly attached to stones or the leaves of water plants. They hatch into perfect little snails which can be seen developing inside the eggs. Snails are hermaphrodite so, during mating, one snail climbs onto the shell of another, thereby establishing the role of male but it later assumes the female role if a third snail climbs onto it.

Males are not always essential. Aphids or Plant lice may reproduce all summer without a male to be seen. They give birth to live daughters. One can be seen emerging while the growing brood accumulates around the mother.

Life history of the Aphid. In summer wingless females (1) reproduce parthenogenetically to produce more wingless females (2). In autumn winged females (3) and males (4) are produced. The winged females produce a wingless form (5) that reproduces sexually with the females which lay eggs (6) that overwinter before starting the cycle again.

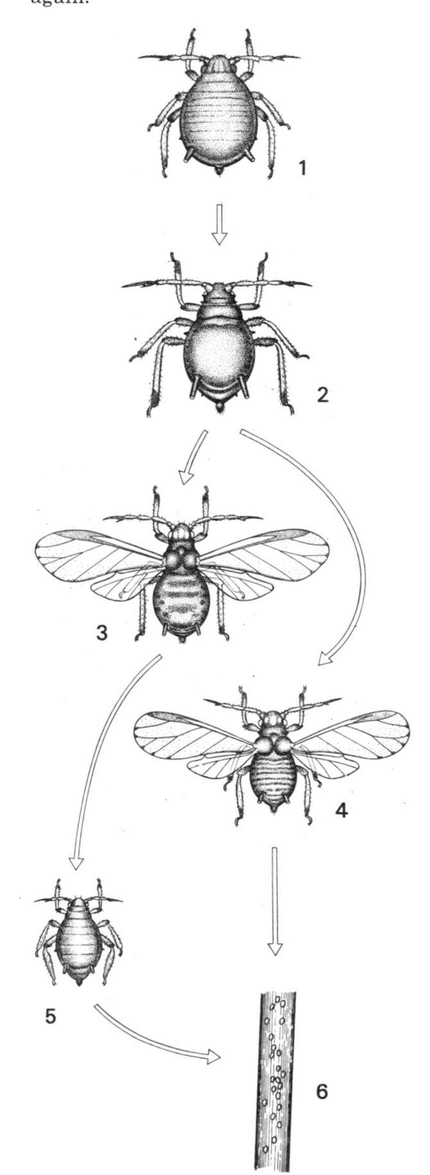

carry half a dozen of these tiny males on almost any any part of her body.

Virgin birth. Reduction of the male animal takes place typically among immobile or very scattered animals but, at the other end of the scale, very numerous animals sometimes dispense with males altogether. It is occasionally possible, even among higher animals, for ova to develop into complete animals without fertilization. We have seen that a sperm has two functions: to stimulate the ovum and to contribute genetic material. The first can be artificially contrived by pricking the ovum with a needle or subjecting it to heat or chemical treatment. In most cases the embryo is misshapen and soon dies, but not always so. There have even been a few mature male turkeys reared from unfertilized eggs.

Reproduction without fertilization is called parthenogenesis, a word derived from the Greek and meaning virgin birth. As fertilization involves doubling of the chromosome pairs parthenogenesis has to proceed either by the ovum's having double chromosomes initially, or by two cells fusing and doubling the chromosomes early in the development of the embryo.

Among the vertebrate animals, parthenogenesis appears as a rare anomaly. In mammals there is no convincing evidence for it although an abortive parthenogenesis is a normal feature of a few species. There are two lizards, the Chequered whiptail lizard of Texas and *Lacerta saxicola* of the Caucasus, in which large-scale collecting has revealed that certain areas have populations in which there is not a single male. Elsewhere there are normal bisexual populations, so the existence of the parthenogenetic populations presents a puzzle. Among some invertebrates of fresh waters and the sea, parthenogenesis is an established annual occurrence in some species of insects, crustaceans, rotifers, nematode worms and molluscs. The most familiar example is found in the aphids or Plant lice.

Several species of these sap-sucking bugs are pests of cultivated plants. During the summer, the population of the Black bean aphid consists solely of wingless females that produce live young continuously and parthenogenetically. They spend their lives sucking sap from leaves and stems of beans, docks and beet while, from the other end of the body, a constant stream of young aphids is being born alive. In warm weather, a single aphid bears as many as 25 daughters in one day and, as these mature in little more than one week, the extended family, or clone, of a single aphid reaches phenomenal numbers. Or they would, if it were not for the birds, lacewings, ladybirds and spiders that eat aphids in huge quantities. In autumn, as temperatures drop, winged, parthenogenetically-produced females, fly to Spindle trees and lay eggs that hatch into males and females. Mating now takes place between these and the fertilized females lay eggs which hatch in the following spring as new 'stem mothers' which colonize new crops and set the cycle in motion again.

Virgin birth is a mechanism that allows aphids to make the best use of the summer abundance of food. Each aphid becomes an assembly line for turning out more aphids. By tapping a plentiful supply of sap, the aphid does not have to waste time and energy searching for food. Neither does it have to wait for a mate to find and court it before reproduction can begin. But the drawback to parthenogenesis is that each aphid is genetically identical to its mother and its thousands of sisters, cousins and daughters and all their issue. They are consequently very vulnerable to changing circumstance. Adverse conditions in the form of a new predator or a change in climate sufficient to kill one aphid will wipe out the whole clone. If they had been

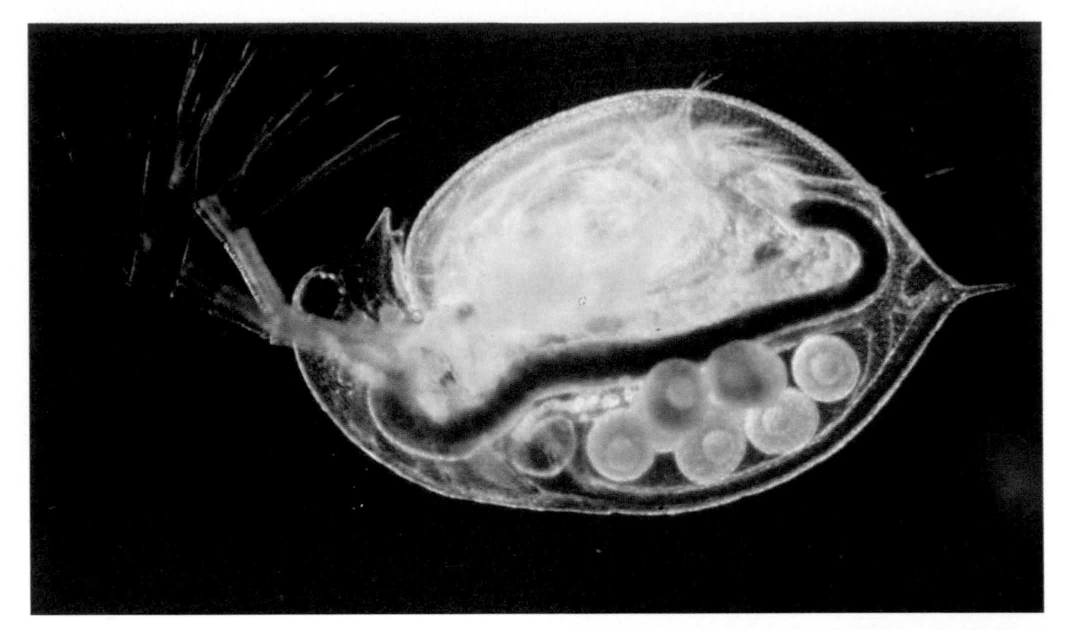

The Water flea *Daphnia* is another animal that reproduces by parthenogenesis or virgin birth. Females lay batches of eggs which are retained in the mother's shell (LEFT) until the baby Water fleas hatch out (BELOW). The babies are all female until conditions become overcrowded. Then, males appear and normal sexual processes ensue.

produced by sexual reproduction, there would have been variation in the population and some might have survived to perpetuate their advantageous characters. Aphids can take this risk during parthenogenetic periods because at the end of the summer sexual reproduction introduces variations into the population so that natural selection can work during the succeeding season by weeding out unsuitable clones.

Another animal that uses parthenogenesis for an explosive increase is the Water flea *Daphnia*. This is a very common crustacean of ponds and lakes, named 'flea' because it is continually leaping up and down in the water with jerking beats of its long antennae. During the summer, when the microscopic algae on which Water fleas feed are blooming (that is, increasing rapidly in numbers), parthenogenetic breeding takes place. Females lay batches of 30 eggs at two or three day intervals and will lay up to 20 such batches in a lifetime. The eggs are retained in the transparent two-sided shell of the female Water flea, that looks like a gaping clam shell, until the Water fleas hatch. The next batch is laid about an hour later. All these are unfertilized eggs. When environmental conditions deteriorate through overcrowding or a shortage of food, however, some of the eggs hatch as males. The male Water fleas fertilize the females and they then lay a different kind of egg, one enclosed in a tough coat that can withstand winter freezing or summer drought.

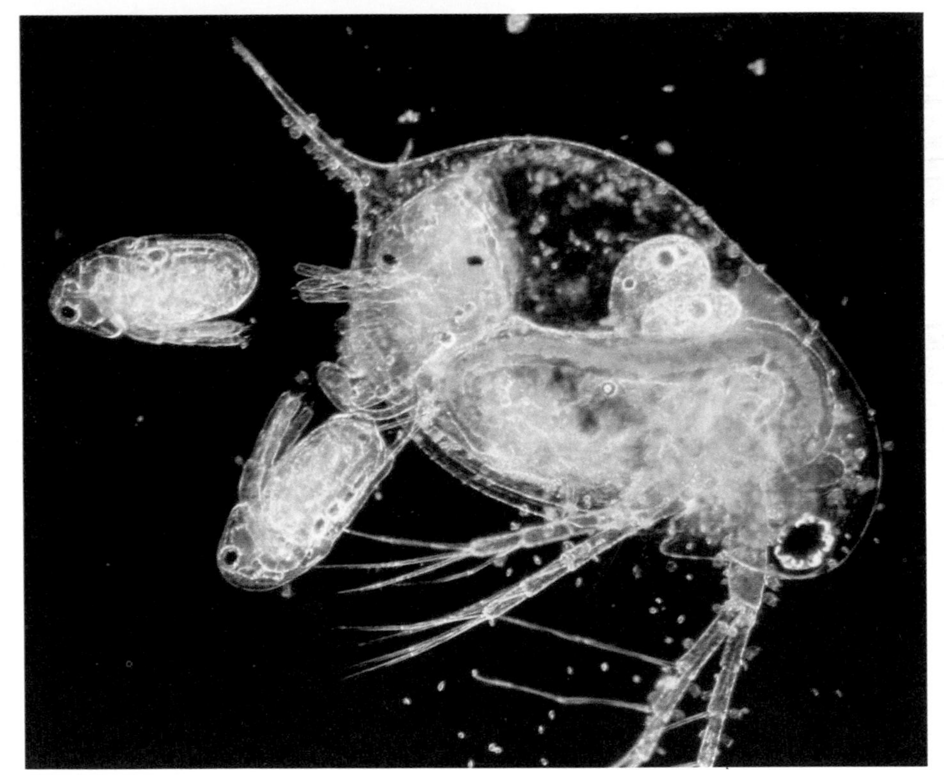

Virgin birth of a different kind is practised by honeybees and ants. The queen is fertilized once in her lifetime, during the nuptial flight when she is chased by the drones, and mates with one of them. His sperms are stored in the queen's body and are used as required. When she first establishes her new colony, the queen lays fertilized eggs which develop into worker bees. These are sterile females and new queens are produced only later in the season by feeding the larvae that hatched from some fertilized eggs on a richer diet. At about the same time, the queen lays some unfertilized, parthenogenetic eggs. These develop into male drones.

Males are, therefore, very rare among the social insects, and also among their relatives the Gall wasps which lead solitary lives, it is unusual to find a male. The Robin's pincushion or Rose bedeguar is a gall covered with a tangle of long branching hairs and it can be 3 in across. In winter it contains the pupa of the Gall wasp and the adult emerges in spring. Unlike the prolific aphids, there is

RIGHT: The queen honeybee is fertilized once in her lifetime, during the nuptial flight. Sperms received during this single mating are stored in her body and used to fertilize each egg before it is laid into the cell awaiting it. Throughout her adult life, the queen is attended by her workers, sterile daughters that hatched from fertilized eggs. The males, or drones, hatch from unfertilized eggs.

Swarming for ants and bees is a free-for-all. Many more males than females are produced so there is intense competition among the males. Somewhere in the mass is the virgin queen ant. Some males become decoys. Scent from the queen rubs onto them and they are chased by other males.

Unlike the male ant or bee, the king termite survives swarming. He stays with his consort and they set up the nest together. Here two pairs have landed, cast their wings and are digging into the ground together. The queen becomes an egg-laying factory but if she dies the colony survives through the activities of secondary and less productive females.

only one generation each year and there is only one male to every 100 females. The males are not prodigious Bluebeards that must mate with dozens of females; most of the females reproduce parthenogenetically. Another familiar gall is the Oak apple. Oak apples grow in spring and each contains several larvae of the Gall wasp. Eventually male and female wasps appear, as they do in aphids. These mate, and eggs are laid in oak roots. Larvae hatch from these in autumn and they make a second kind of gall in which they feed until the following summer when they emerge as adults. This generation is entirely of wingless females and they lay eggs parthenogenetically. Larvae from these eggs form Oak apples again, so there is a biannual cycle of two generations, one bisexual and the other parthenogenetic. There is a similar alternation of generations in the Oak marble gall wasp, except that the

Oak apple galls grow from buds containing the eggs of a Gall wasp. These hatch and the mature gall contains several grubs from which adult wasps of both sexes develop. The females lay their eggs on oak roots and the resulting larvae form a second kind of gall and develop into parthogenetic wingless adults. They emerge one year later and crawl up the oak trunk and, without being fertilized, lay eggs which give rise to the next generation of oak apple larvae.

Telenomus is a minute Chalcid wasp that lays its eggs in the eggs of other insects. Each egg divides repeatedly and many adult wasps eventually emerge where a single egg was laid. About 200 wasps emerged from each of these Drinker moth eggs.

parthenogenetic females are winged and migrate from Common oaks to Turkey oaks.

Child brides. What the advantage of this two-generation, part-parthenogenetic system might be is anyone's guess. It does not seem to have the advantage of the massive population increase so often found in fully parthenogenetic animals like the aphids and in tapeworms and other parasites with the complicated life-cycles that make it a necessity. However, another group of gall-making insects provides an example of another method of speeding up the production of offspring. The Gall midges are small flies which look like mosquitoes. There are many species and most groups of plants throughout the world are attacked by their own Gall midges. Some of these midges start to breed before they have grown up. This is called paedogenesis from the Greek word *paidos* – of a child. The adult female lays several large parthenogenetic eggs which hatch into larvae. Each larva then produces precocious eggs that develop into larvae and devour the parent from inside. They in their turn produce eggs that grow into larvae and these also devour their parents. This can continue for several generations without a single adult. Eventually all burrow out of the original mother leaving her an empty husk. After several generations the larvae pupate and turn into males and females.

Reproduction at an even earlier stage of the life-cycle is called polyembryony. In certain Chalcid wasps the eggs themselves reproduce. The Chalcid wasps are parasites so minute that some can lay eggs in the eggs of moths. When the moth egg hatches the wasp egg is lodged in the caterpillar's body. The egg divides into a mass of cells which organize themselves into a number of embryos, each of which results in an adult wasp that eventually burrows out of the now-dead caterpillar. One Chalcid wasp has been recorded as producing 1,000 wasps from a single egg. And this is not all. Some chalcids make sure that one egg only is laid in each host egg but others are not so exclusive and several eggs may be laid in a single host. More than 3,000 wasps have been counted as they emerged from one caterpillar.

Paedogenesis and polyembryony are two more instances of the incredible variety in the habits and strategies of insects. No other group of animals has found such an assortment of solutions to the problems of life. They seem to have combined all the life styles found in other groups of the animal kingdom together with some that are uniquely their own. Paedogenesis is one life style that is used by other animals but, by some quirk of nomenclature, the phenomenon is then known as neoteny. The most familiar example of neoteny is the familiar classroom exhibit – the axolotl. This is the larva or tadpole of a salamander. Adult salamanders, like the related newts, normally have an aquatic tadpole which undergoes a change of body plan or metamorphosis to become the land-living sexually mature adult form. In the axolotl, a native of

RIGHT: A common aquarium pet with a strange life story, the axolotl is a Mexican amphibian. It is an outsize larva, like the larva of a newt, and retains the feathery external gills. The axolotl becomes sexually mature and reproduces while still in this physically immature condition. On metamorphosis it becomes a salamander.

A normal adult salamander has lungs and a slender tail, like an adult newt. These European Fire salamanders were thought, in mediaeval times, to be able to survive fire. The female gives birth to live larvae, having retained the eggs in her body.

Mexico, the tadpole becomes sexually mature without metamorphosing. It retains the larval feathery external gills and broad fin running down the back and the tail.

In the wild, axolotls are confined to certain lakes around Mexico City. *Axolotl* is Mexican for 'water sport' and axolotls are caught and roasted, being regarded as delicacies. The full explanation for neoteny is more complicated and the process itself is not fully understood except that it is caused by a deficiency in the diet that prevents the manufacture of thyroxine by the thyroid gland. Thyroxine is the hormone that controls metamorphosis. As an essential ingredient in thyroxine is iodine, the shortage of iodine in water is a very likely cause for the axolotl's failure to metamorphose. Certainly, the addition of iodine to the water in an aquarium can cause axolotls to change into salamanders.

There are some amphibians which are permanently neotenous. The amphiuma, mudpuppy and sirens of the United States differ from the axolotl in that they cannot be persuaded to metamorphose under any circumstances. The amphiuma does, however, have a partial metamorphosis. It develops lungs but retains one pair of the original three pairs of gill slits and it lays its eggs on land. Little is known of the courtship of these animals in their natural environment. They are boringly inactive when kept in aquaria except that the courtship of amphiumas is surprisingly violent. During the summer months fresh mutilations appear on their bodies, due to limbs being torn off during rough courtship gambols.

Intersexes and sex changes. As a result of biological 'mistakes' and other abnormal circumstances, animals sometimes show a mixture of characters of both sexes. This is a pathological condition and is not the same as true hermaphroditism in which an animal functions properly as both male and female. Not surprisingly, the insects, those masters of the bizarre, furnish examples of intersexes. The Gypsy moth sometimes occurs as a functioning male even when its genetic character is female but a variety of mixtures of male and female can be obtained by exposing the moths to high temperatures. Intersexes of the Human louse appear when the head louse is crossed with the body louse. The two kinds usually stay within their respective body regions and do not meet and interbreed. If, however, they are brought together and induced to breed, as many as a fifth of the progeny turn out as masculinized females, with both testes and ovaries.

Parasitization is a frequent cause of partial sex change. The common chronomid midges are parasitized by a nematode worm. The worm destroys the midge's ovaries just before it pupates. The secondary sexual characters, such as the lack of hairs on antennae and forelegs which characterize females, are retained but testes and a complete set of male reproductive organs are formed in place of the destroyed ovaries.

A change of sex can be part of an animal's regular pattern of reproduction, as in the Slipper limpet where it is a mechanism that ensures that a chain of the limpets always contains individuals of both sexes. Sex-change is also found in certain tropical marine fishes where it forms part of a deeper strategy. There is a wrasse, one of the Cleaner fishes, which lives on the Great Barrier Reef of Australia. Cleaner fishes take up station near a prominent piece of coral or rock and are visited by other fishes who allow the Cleaner fish to remove parasites from the skin and even the inside of the mouth. This alone is a remarkable piece of behaviour but this particular wrasse (*Labroides dimidiatus*) lives in small groups of eight to ten in-

dividuals, each of which has its own territory. Only one of the group is a male, the rest are females. The male is the boss dominating all his females, who in turn are ranked according to size. The dominant female has a territory that is superimposed on the male's. If the male dies, the dominant female not only takes over his territory, she actually takes over his functions. Within a few hours she is behaving like a male and a few days later she (now he) is fertilizing the remaining females. The change is rapid because the females are really hermaphrodites. They have testis tissue in their ovaries which is kept inactive through the influence of the male's dominating behaviour.

The value of this strange system seems to be that the most successful females become top rankers and, once they have achieved a leading position, the way in which they can contribute most to the next generation is to become males and fertilize several other females.

The sex life of Cleaner fishes. The blue, striped Cleaner wrasse lives in small groups of females except that the most senior fish becomes a male, a strange organisation that is described in the text. The yellow Green moon wrasse has two colour forms. Yellow forms mate in a swarm but some individuals continue growing and, if female, change sex. These are called terminal or super males. They become patterned in green, orange and yellow and each mates singly with a yellow female.

SIGNALLING AND SELECTION IN INSECTS

The insects as a group have well organized and often elaborate courtship and mating behaviour. This is a consequence of their perfection of body mechanism. They are well endowed with sense organs and are very active, being one of the few animals to have mastered flying. Good sense organs are essential for an active way of life, as is a well-developed brain, and the two attributes make for an active and, to us, interesting sex life. The million and a half known species of insects have invaded nearly every part of the world and have taken up almost every conceivable way of life, so it is not surprising that their ways of courting, mating and other aspects of reproduction are also extremely diverse. They are one of the most successful of animal groups and this is, at least in part, due to the constraint put on their mating by the demands of internal fertilization.

Insects cannot mate in the free and easy way of marine animals such as Sea urchins and mussels, nor even in the manner of crayfish and crabs. Internal fertilization is essential in a land-living group of animals as is foreshadowed by the woodlice, the sole land dwellers among the crustaceans. Sperms cannot be exposed to the air. They must be placed in the female's body, and eggs must be retained there at least as long as is necessary for a waterproof covering to be made. So, although internal fertilization is an added complexity to life, it is immensely advantageous because it gives the animal considerable independence from its environment. A drawback is that the animal has to expend energy in finding and making close contact with the opposite sex, but insects counter this by the elaboration of signal systems and other energy-conserving strategies.

The pattern of bands and eyespots in male Grayling butterflies that are used to signal to females during courtship.

Dangerous courtship. The close contact and actual linking of the sexes in the act of mating can involve the male of flesh-eating animals in physical danger. Where behaviour is simple and intelligence low, the male must ensure that the female recognizes him for what he is. The male spider signals his sex and intention and thereby saves his life but the male Praying mantis is not so lucky. The Praying mantis is so named from its habit of standing motionless with forelegs raised in attitude of prayer. The posture is far from peaceful; the tibia and femur of the forelegs are armed with sharp spines and snap together like a living gin-trap for seizing their prey in a lightning snatch.

Mantids are quite unselective in their choice of food. They sometimes eat small birds and lizards and they are happy to snap up other mantids, so mating is a hazard for males. On sighting a female, the male freezes, then moves forward ever so slowly, perhaps one inch per five minutes, until he is close enough to leap on the female. Very often he is not quick enough and is himself seized. The female then proceeds to eat his head but, far from being a hindrance, the removal of the head actually stimulates the male's copulating activities through the destruction of an inhibitory part of the brain. And the female benefits from an extra meal which helps towards forming the eggs.

In New Zealand, there is a wingless fly of the family Empidae which shares with the Praying mantis the jack-knife action of the forelegs. The Empidae or Dance flies are carnivorous and the males have adopted stratagems with all the appearance of cunning to overcome the problem of being eaten by their mates. At mating time, the females gather in a sunny place and fly to and fro circling each other, hence the family name. The males search for these swarms but to fly straight in would be like walking into a lion's den. To save its skin the male Dance fly catches an insect, carries it to the group of females and approaches one who grabs the proffered gift. While she is busy devouring the insect, the male grapples with her and flies to a nearby bush where he hangs from a leaf while copulating with her. Some Dance flies go one stage further in their diverting of the female's attention. Members of the genus *Hilara* wrap their victim in strands of silk spun from glands on their forelegs to prevent it struggling and some practise a deception by finding a small petal, a bud scale or a minute piece of mica from the soil, which is then wrapped in silk and presented to the female. She is kept busy investigating the gift-wrapped bundle while mating takes place.

Moths and butterflies. For the majority of insects the problem is that of finding the female. Individuals live separate lives, usually meeting only by chance, so some method is needed actively to bring the sexes together. Among many moths attraction of males by the females is mediated through the sense of smell. It has long been known by moth collectors that a caged virgin female attracts a

In a cannibal fertility rite, this female Praying mantis eats the head of the male while clasping the severed abdomen with one leg. Premature dismemberment does not terminate the male's ultimate purpose. Copulation can proceed despite the loss of his head.

Placating a carnivorous mate is a necessity for some animals. This acrobatic Dance fly hangs from a plant while mating. The attention of his suspended mate is distracted by the gift of a small fly.

bevy of suitors attempting to get at her. When these observations were first made, it seemed too incredible that so small an animal as a moth could trace a scent over a distance of some miles, as the German naturalist, Mell, had demonstrated by marking male silk-moths with nicks in the wings and releasing them at every station on the railway line running from his home. It was considered that some mysterious rays were involved. Of course, this was nonsense but it has often been the practice to invent some unknown sixth sense to explain a phenomenon which is a mystery only because of the inability of our senses to match up to those of the animal under study.

The sex-attractant scent of moths is part of a delightfully simple mechanism. When ready to mate, the female 'calls' by raising the tip of the abdomen and fluttering her wings to drive a current of air past the scent glands on the abdomen. The scent disperses downwind and serves several functions at once. It identifies the species of the moth, it indicates that she is ready to mate and it stimulates the males into activity. Scent is a useful medium for signalling because it penetrates every crevice and stimulates the males wherever they may be. They pick up minute traces of scent on their feathery antennae and immediately fly up-wind. When a male gets close to the female the concentration of scent is sufficient for him to alter course and fly directly to her. The scent has, therefore, informed the male of the position of a receptive female and has also stimulated him to mate with her.

The use of scent has allowed some female moths to abandon flight. The large flight muscles are not needed and the huge energy requirements for flight can be diverted into egg production. The females of some species do not even leave the cocoons in which they pupated. They are fertilized and lay their eggs *in situ*. The moth *Orygia splendida* of southern Europe is unusual in that the male has a scent as well. He is attracted by the female's scent to where she is lying in her cocoon and smears his own scent over it. On receiving his scent, the female tears a hole in the cocoon with one of her claws so that the male can get in to mate with her. The eggs are laid in the cocoon and the female dies alongside them. The Gypsy moth is a more familiar member of the same family. It has been introduced from Europe to the United States where it soon became a pest. To combat the moth, its sex-attractant scent has been identified and an artificial substitute called 'gyplure' has been manufactured to bait traps for the males to prevent their mating with the females.

The use of the sense of smell seems to be

Feathery antennae of the male Luna moth act as nets to trap scent particles wafting from the female. The scent both stimulates the male to search for the female and guides his search.

After emerging from the pupa, the female Luna moth waits passively, relying on her scent to draw suitors to her.

universal in the courtship of moths, which perhaps is not surprising as moths are active mainly by night. In the day-flying butterflies, sight is used for bringing the sexes together and scent is used mainly for stimulating the mating act. The courtship ritual of the grayling has been observed closely by Niko Tinbergen and his Dutch colleagues. Graylings emerge from the chrysalis in July and spend much of their time feeding at blackberry, thistle and other flowers of the hedgerows and on heather flowers in open country.

As the urge to breed overtakes the male grayling, he stops feeding and takes up position on the ground. Whenever another butterfly flies over, he flies up to investigate. If the butterfly is a virgin female grayling, a set pattern of behaviour unfolds. She reacts to the male's approach by settling on the ground. The male alights behind her and walks round to face her. If the female is ready to mate she remains still and the male starts his courtship. He jerks his wings upright and opens and closes the front wings to show off the patterning of bands and eyespots. After waving his wings in this manner for some seconds he spreads his forewings, then, quivering, he appears to bow and closes his

wings, trapping the female's antennae between them. So far, courtship has been visual but the female's antennae are now pressed against patches of perfumed scales on the male's forewings. The scent from the scales is a pheromone that gives the final stimulus for the female to receive the male. He releases the female's antennae and walks round her until standing directly behind her. Then he couples with her.

Percussion and scraping. Among the most primitive insects are the wingless springtails which have changed little for millions of years. Small and inconspicuous, the springtails usually go unnoticed, but their courtship was described as early as 1871. The male and female were said to 'butt one another, standing face to face and moving backwards and forwards like two playful lambs'. What might be termed a corporal stimulus is also applied by Hover flies which hover on rapidly humming wings and come together head-on, swinging to and fro like the weights on two pendulums. Sometimes the male gets misled and butts a bee by mistake. A sort of head-butting-at-distance is used by the Death-

RIGHT: Common blue butterflies mate on a stem. The sperms are transferred in a packet or spermatophore.

LEFT: Sexual variety in insects. Instead of the male mounting the female's back as in so many animals, these Amati moths copulate tail-to-tail. Mating may last for over an hour.

A cluster of ladybirds gathered for mating. The females depart to lay their eggs on a plant infected by aphids, so that emerging larvae have a plentiful supply of food.

LEFT: Two Rhinoceros beetles meet in a challenge. The horn is a thickened extension of the body cuticle and the two males push against each other in rivalry until one retires.

watch beetle which can be heard tapping in old beams at night, so giving rise to the idea that the life of one of the household is rapidly ticking away. The sound is nothing more sinister than the beetle banging its head against the walls of its burrow as a mating call. Although we can hear the Death-watch beetle ticking, the beetles are without organs of hearing and feel the vibrations coming through the timber.

Head butting is an unsophisticated stimulus requiring no sense organs other than those registering impact or vibration and cannot convey much information. As soon as a specialized organ for detecting vibrations is developed the field is opened for an order of increase in communication. Such an organ is the ear which detects sounds, or airborne vibrations. It is regarded by scientists as preferable to refer to 'hearing organs' rather than to ears because the insect's organ works on a different principle from our ears. This seems to be an unnecessary technicality as an insect's eyes are very different from ours, yet no one talks of an insect's 'seeing organs'.

Sound is important in the communication of courtship signals of the cicadas and the crickets and grasshoppers, two groups of insects which have evolved both efficient organs of hearing and structures for the production of sound. In the same way as insects have hearing organs rather than ears, so they cannot be said to have proper voices because the sounds are not made by the passage of air over over vocal chords. The 'voices' of these insects are generated mechanically. What they are producing is in effect instrumental music.

The cicadas are a family of bugs whose shrill, piercing calls are so much a part of the atmosphere of warmer countries. The loudest cicadas can be heard over a distance of $\frac{1}{4}$ mile and, near at hand, they are deafening. The noise is made by a pair of tymbals on the abdomen. These are stretched membranes like the skin of a drum which are pulled in by special muscles and allowed to 'click' back. By 'clicking' at a rate of 100–500 times a second, the whine of a cicada is produced.

The crickets and grasshoppers produce a softer, quieter sound but, nevertheless, a chorus can still be quite maddening. These jumping insects which, according to fable, pass the summer singing without a care, are classified into Short-horned grasshoppers – family Acrididae, Long-horned grasshoppers or bush-crickets – family Tettigonidae, and the crickets – family Gryllidae, all distinguished by the length of the antennae. Their 'songs' are produced by rubbing parts of the body together, a process called stridulation. Short-horned grasshoppers rub a row of knobs on the femur of the hindleg against a vein on the forewing to set it vibrating, while crickets and bush-crickets rub the two forewings together. Each pulse of sound, of which there are one or more to each 'chirp', is made by one full stroke of forewing or femur.

These sound-making insects differ in one important respect from the butterflies and moths because the males signal to the females rather than the other way round. The females have to take the initiative and, if willing to mate, seek out the males. The 'song' is a signature tune which identifies the male to the females by its pattern of chirps, which is extremely specific and is the sole means of

Male and female cicadas are brought together by the deafening and monotonous calls of the male. The female is usually mute but some species are vocal in both sexes. The sound is made by the vibration of a membrane stretched across an aperture like the parchment of a drum.

The female Bush cricket or Long-horned grasshopper is attracted to the male by his 'song'. He 'sings' or stridulates by rubbing the roughened edge of the right forewing against the smooth left forewing. The ears are on the forelegs, just by the 'knee' joint.

keeping the species from interbreeding. A female grasshopper can be lured to a male of another species and induced to mate with him if he is prevented from stridulating and taped songs of a male of her own species are played over a loudspeaker.

Grasshoppers and crickets of species so similar that they are difficult to tell apart on appearance alone can be distinguished by sound. In the eastern United States there are species so similar that their differences in appearance have only been found after they have been distinguished by comparison of songs. Two American species, *Nemobius allardi* and *N. tinnulus*, are distinguished chorally because the chirping of the former is three per second faster. This is sufficient to prevent interbreeding and, of all the crickets and grasshoppers in the eastern United States, only two crickets, *Acheta pennsylvanicus* and *A. veletis*, have identical 'songs'. Although these two species live in the same places, they do not mix because the former matures and sings in the autumn while the latter is active in spring.

The sophistication of the stridulating mechanism is shown by the specific differences in courtship songs and also in the

Short-horned grasshoppers stridulate by rubbing the femur of the hindleg against the forewing. Females are attracted from a distance to the singing male, who then begins a more active courtship.

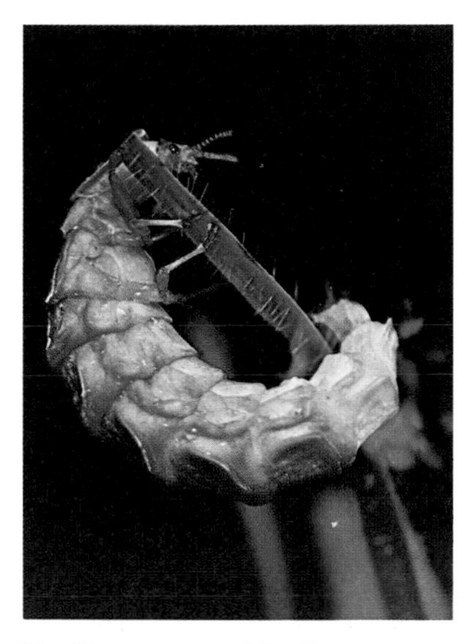

The Glow worms and fireflies are beetles which seldom fly. The females often have minute wings or no wings at all. At night both sexes flash a green or red identification signal.

repertoire of each species. The Field cricket *Teleogryllus commodus* has the most varied repertoire known. Its song contains two rhythms, one for serenading and one for haranguing other males, so as to keep them spaced out. Once a female has been attracted to a male, he uses a courtship song to bring her into the mating position. There is also an aggressive call, a courtship interruption call, a post-copulatory call and a recognition call. An aggressive call is used in fighting other males.

Mating swarms. A chorus of grasshoppers, cicadas or crickets makes the finding of a mate so much the easier for the females. It also gives them more of a choice, but the importance for swarming in insects is to make a trysting place where one sex can meet the other; the dance-hall in human idiom. One of the more spectacular assemblies of insects, unfortunately fast disappearing from built-up areas, is that of fireflies. In tropical countries fireflies sometimes put on a most spectacular show of illuminations.

Most fireflies do not form aggregations. Flashing is used as a signal between pairs of fireflies. The flash is produced by the oxidation of a complex substance called luciferin which, unlike an electric light bulb, gives out light with very little output of heat. The light is produced in a light organ backed by a reflector and covered with a lens of transparent cuticle. Each firefly delivers about 1/40 candlepower but the wavelength of this light is that to which human eyes are particularly sensitive so even the gentle glow of a few fireflies is sufficient to read by and firefly lanterns were once common in the Orient. The male firefly flashes to attract the female and she replies to identify herself to him. One firefly of North America, where they are often called lightning bugs, flashes every 5.8 seconds, an extremely accurate time-interval between flashes being a feature of fireflies. The female replies exactly two seconds later to provide the necessary identification. The male firefly approaches any light that flashes at this particular interval and synchronizes with it so that eventually a mass co-ordinated flashing builds up.

In the Orient, the night may become filled with flashing as on a river bank outside Bangkok, described by H. M. Smith in 1935:

'Imagine a tree 35–40 ft high thickly covered with small ovate leaves, apparently with a firefly on every leaf and all the fireflies flashing in perfect unison at the rate of about three times in every two seconds, the tree being in complete darkness between the flashes. . . . Imagine a tenth of a mile of riverfront with an unbroken line of *Sonneratia* trees with fireflies on every leaf flashing in unison. . . . Then, if one's imagination is sufficiently vivid, he may form some conception of this amazing spectacle'. The spectacle continued all night for weeks on end.

Quite why these fireflies gather to flash in unison is not known except that it would appear to be a mechanism to bring all the fireflies together for mating and this is a reasonable explanation of the swarming of a great many insects. Many of these swarmings are familiar country sights. There are the dancing swarms of the Winter gnat which can be seen on warm, calm winter days and midges are similarly seen dancing on summer evenings. Rivers and lakes are the scenes of mating dances of mayflies, insects unique in having two winged stages. After a long life in the water the nymph crawls out of the water and splits open to reveal the winged subimago, the 'dun' of the flyfisher, whose wings are dulled by a fine pile of microscopic hairs. After a few hours the skin is shed again and the shining imago or full adult emerges.

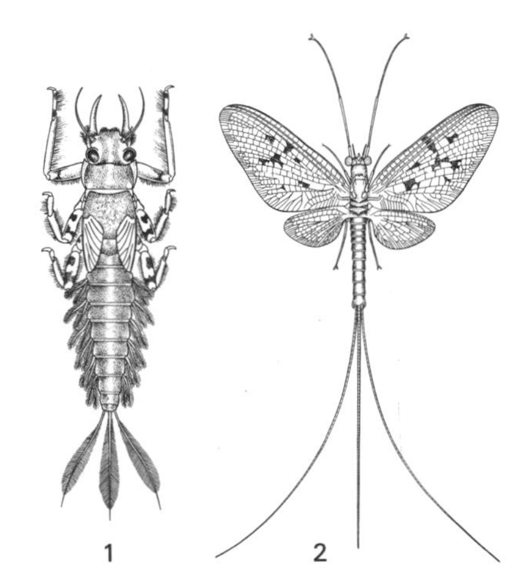

The aquatic nymph (1) of a mayfly and the adult (2) that swarms to mate and sometimes lives for only a few hours.

Three pairs of the American damselfly *Argia moesta*. Mating has already taken place but the males continue to grasp their mates with claspers on the tip of the abdomen. Meanwhile, the females settle and search with the tip of the abdomen for a suitable place on an aquatic plant in which to lay their eggs.

Mayflies usually emerge in a swarm, the 'hatch' of the fisherman, and the surface of the water becomes covered with delicate insects fluttering up and falling. The imagos live only a few hours, or at most a few days. Their sole function is to mate and lay eggs.

Swarms of mayflies, midges and mosquitoes consist mainly of males which dance together as an advertisement to the females. As soon as a mosquito female enters the swarm she is grabbed by one of the males and carried away from the crowd to mate without disturbance. Although the female finds the swarms by sight, her signal to the males is auditory. They are attracted by the buzzing of her wings from a range of about 10 inches. The hearing organs of the male mosquito are in its antennae. These are feathery and, by analogy with the antennae of the male moth, might be expected to bear organs of smell. However, they are set in vibration by sound waves to stimulate sense organs at their bases. Male Yellow-fever mosquitoes are attracted by sounds having a frequency of 300–800 cycles per second. The wingbeats of the females emit sounds at 450–600 cps, thus falling within the range to which males are sensitive. Moreover, the system has additional refinements in that the feathery antennae do not unfold until a male mosquito becomes sexually mature and immature females have slow wingbeats whose low-pitched buzzing does not attract males. So time is not wasted on courtship by immature mosquitoes.

The male Yellow-fever mosquito undergoes a remarkable change at maturity. Its rear end rotates through 180 degrees and until this happens it is unable to mate. This rotation occurs in the males of all true flies. The effect is to change the position of two pairs of hooked clasping organs so that they can engage and hold the tip of the female's abdomen during mating. Mating is very brief and lasts only 15–20 seconds. The male grasps the female's body with his hooked legs and swings under her as they fall to the ground. The tips of their abdomens come together, the claspers lock on and the male's aedagus, or penis, is forced into the female and also locks in place. Sperms are instantly discharged and the pair separate.

Guaranteeing children. The biological function of every animal, both male and female, is to leave as many offspring as possible and the destiny of the species hinges on the best, most healthy and 'fittest', individuals leaving the most offspring. It is on this basis that natural selection works. The female's contribution is to ensure that she is fertilized and lays as many eggs or rears as many young as possible. For males, the necessity is to fertilize as many females as possible, thereby ensuring as great a genetic contribution to the next generation as possible. As they are capable of mating several times, there is competition among males to fertilize the available females. How a male ensures that the next generation will include his own progeny is called the reproductive strategy. This is a term that will assume an important place in the latter half of this book. The evolution of mating from the broadcast scattering of gametes to internal fertilization after a courtship ceremony has allowed males to determine their siring of offspring. As the female can reproduce after mating with one

RIGHT: Another two pairs of damselflies set about egg-laying. This species prefers to lay well below the water surface and the female climbs down the plant stem until completely submerged. The male either submerges with her or breaks free.

male, male animals have evolved means of making sure that the sperms of one male alone shall be responsible for fertilization and there is competition to be that one male. The result is sexual selection.

Insects have evolved the same reproductive strategies that are to be found in the vertebrate animals. We have already seen one strategy employed by Dance flies and mosquitoes. When a male has secured a female, the pair drop out of the swarm and away from competitors. Provided that the female does not return to the swarm, there is little likelihood of another male inseminating her and fathering some of her progeny.

Another strategy is physically to guard the female so that no other male can mount her. Thus copulation often continues for a considerable time after insemination. Houseflies copulate for one hour but the sperms are transferred in the first 10 minutes. Once inseminated, the female loses her receptivity and reacts aggressively to would-be suitors. The change in her behaviour is caused by substances in the male's seminal fluid and he stays mounted until this has taken place.

Guarding the female is the reason for the odd-looking tandem-flights of dragonflies and damselflies. Their method of mating is quite unusual. With few preliminaries, the male alights on the back of the female and curls his abdomen forward to grasp her head or thorax with the pair of claspers at the tip of his abdomen. He then releases the hold with his legs and flies linked in tandem with the female. But before coupling takes place, the male will have performed an unusual action. He will have transferred his sperms from his genital organs on the tip of the abdomen to an accessory organ farther forward on the abdomen. Later, when the pair are flying in tandem the female curls her abdomen to make contact with the accessory organ and receive the sperms. The pair continue flying in tandem until the female lands on water plants to lay her eggs in their tissues or flies low over the water and sheds them so that they sink to the bottom.

Flying in tandem guards the female from other males and, in some dragonflies, the male detaches himself and flies alongside, darting off to drive away intruders. Guarding the female from the attentions of other males is only part of the dragonflies' reproductive strategy. There is competition among the males to secure the females. After they have spent several days feeding, the males take up territories, areas over the stretches of water where the eggs will be laid, and guard them against other males. Fights sometimes break out and the older males show signs of wear in the form of tattered wings and missing legs. As there is only a limited space to be divided into territories, only a certain number of males can establish themselves. The rest have to wait until the territory holders die. Only territory holders can secure females, so there is an intrasexual competition among males resulting in a limited number of males breeding and these are the 'fittest' available.

The competition between males is heightened by a dominance hierarchy in which the males establish an order of social rank. The fights determine which are the strongest or dominant males. They take part in more matings than their weaker, subordinate colleagues, so there is an additional system whereby the 'best' and 'fittest' males, the ones that are strongest, are most likely to father the next generation. The two mechanisms for putting this into effect, namely territorial behaviour and establishing a peck order of rank, are to be found in reproductive strategies in species throughout the 'higher animals' and form an important theme in the remaining chapters of this book.

Before mating takes place, the male dragonfly or damselfly transfers his sperms to a storage organ at the base of the abdomen. He then seizes a female with the claspers at the tip of his abdomen and she curls her abdomen forward to receive the sperms from him.

COLD-BLOODED COURTSHIP

Apart from an excursion into the strange world of hermaphrodites, virgin birth and intersexes, we have been tracing the development of sexual reproduction among the invertebrate animals from the broadcast, random scattering of gametes into water to sophisticated methods of courtship and internal fertilization, as found in the insects. Courtship and internal fertilization are the keys to both life on land and the evolution of reproductive strategies by which individuals try to leave as many offspring as they can. Similar trends can be discerned among the vertebrate or backboned animals, and they often involve the same or similar mechanisms as are found in insects and other invertebrates. In the sequence of fishes, amphibians and reptiles followed in this chapter we can see how an increasing independence of water is matched by the adoption of internal fertilization.

Like other animal groups, the first vertebrates lived in the sea. We do not know much about these early vertebrates because they disappeared many millions of years ago and have left little trace, in the form of fossils, of what they looked like and even less of how they lived. We can be sure that they were simple animals with little specialization of sense organs, and so we can deduce that their mating habits and reproductive machinery must have been similarly simple.

Reliable knowledge of vertebrate mating habits therefore starts with the fishes, an abundant, widespread and very diverse group of animals. Basically, there are two main groups of fishes: the cartilaginous fishes, which include sharks and rays, and the true or bony fishes. In the former, the skeleton is made of a relatively soft cartilage and not of hard bone as in bony fishes and other vertebrates.

Internal fertilization is the rule among cartilaginous fishes but most bony fishes practise external fertilization. However courtship in bony fishes brings the sexes so close that the act of mating ensures that there is little wastage of gametes and that a female is fertilized by the particular male which has courted her. The amphibians (salamanders, newts, frogs and toads) are descended from bony fishes that learned to creep about on dry land but they are tied to water for reproduction, because fertilization is still external and the eggs and young need to develop in water. There are trends towards internal fertilization in some amphibians but the breakthrough comes with the reptiles whose great advance over the amphibians is that they are independent of water. They practise internal fertilization and this enables their eggs to be covered with a waterproof shell before they are laid.

The courtship of fishes has been a favourite subject for zoologists studying the origins and mechanisms of animal behaviour because the smaller species of fishes are easy to keep and observe within the confines of an aquarium and their behaviour is extremely simple and rigid. Fish courtship is very much like a mime or masque in which not a word is spoken but the 'dialogue' is made quite clear by gestures, facial expressions and expressive movements of the body. A conversation between human beings is basically an interchange of meaningful sounds whereas that of fishes is an interchange of gestures. Consequently it is easy to classify and define each part of a fish's conversation and to analyze its meaning and significance.

Cichlids and sticklebacks. The cichlid fishes, including small tropical fishes such as the angelfishes, the Orange chromide, the jewelfish and many kinds of *Tilapia* are popular with aquarists because of their bright colours and the ease with which they breed in

The many species of cichlids are popular aquarium fishes which breed readily in captivity. The males establish territories from which they expel rivals, while females are courted with much chasing; first aggressively, then seductively.

captivity. Some of them formed the subjects of a classic story of biological research in which the Dutch biologists, G.P. Baerends and J. M. Baerends-van Roon made a minute analysis of their behaviour and found that the 'gestures' making up their behaviour are extremely stereotyped movements that show exactly what the fish will do next. They form a simple language whose advantage is that it is quite clear and unequivocal.

When two male cichlids meet their motivation is quite clear, both to each other and to an informed human observer. They swim round each other then turn to face head-on with gill-covers raised, fins spread and mouth open to show a coloured throat. For all the world, they are like two mediaeval armies confronting each other with banners waving, lances couched and challenges shouted. In the same way as a human challenge consists of exaggerated strutting and arm waving, so the cichlid's threat is an exaggeration of normal body movements. The gill-cover raising and gaping are exaggerated breathing movements. Then, like Chinese armies of old, if one fish decides it is likely to lose the contest, its challenge changes to submission, without a blow being struck. The mouth and gill-covers are closed and the fins are furled. The colours become subdued and the cichlid sneaks away.

The reaction of the female cichlid to the male's threat display is to turn away his aggression by showing submission but, instead of retreating, she holds her ground and dodges his attacks. Gradually the male realizes that she is female and not a rival male. He becomes less aggressive and the message changes as he tries to woo her. He is telling her to be prepared for spawning. So, by simple movements, cichlid fishes establish each other's sex and courtship can proceed.

Another fish in which courtship has been exhaustively studied is the Three-spined stickleback. It demonstrates as clearly as anything could the interplay of stereotyped gestures between male and female to bring the pair progressively from first meeting to fertilization of the eggs. So rigid is the sequence that one stickleback cannot perform a particular gesture until it gets the 'go-ahead' in the form of the correct gesture from the other.

As the water of ponds and streams warms in the springtime sun, the sticklebacks move into shallow water, the schools break up and they come into breeding condition, which, in the male, is indicated by the belly turning red and the eyes turning blue. He sets the scene for courtship and rearing the family by establishing a territory and making a nest out of pieces of waterplant glued together with secretions from his kidneys.

While the male is preparing the nest other males will be trespassing as they, too, are trying to find somewhere to set up home. The owner immediately recognizes the intruder as a male stickleback because of his red belly. The red colour is the sole signal. The resident will not, for example, attack a perfect model

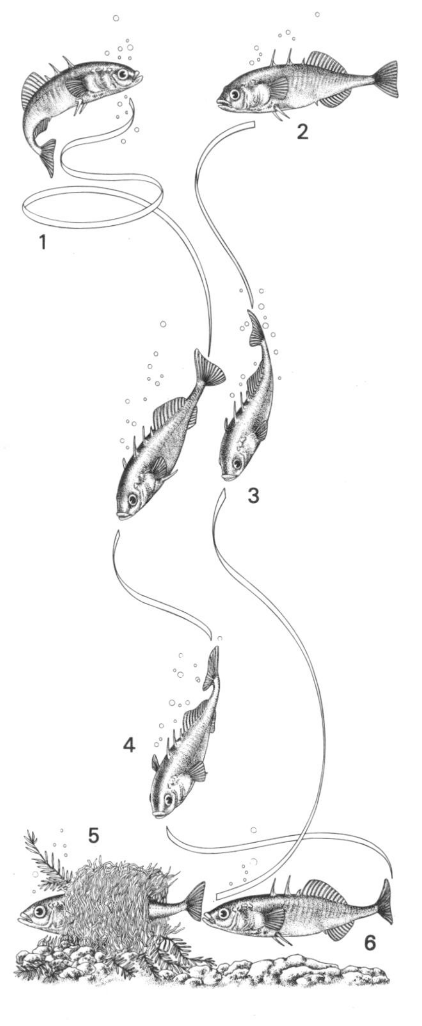

LEFT: Solicitious parenthood by a pair of angelfish. Members of the cichlid family, angelfish form a firm partnership until their young can fend for themselves. They prepare a site for spawning by clearing the surface of a leaf. After the eggs have been laid, the parents fan them continuously.

The female Three-spined stickleback plays a minor role in breeding. The male, distinguished by his red belly, builds a nest of waterweeds glued together with secretions from his kidneys (TOP). When a female approaches, he comes out to court her (LEFT) and (RIGHT) shows her the nest. After the eggs have been laid, the female departs and the male looks after them.

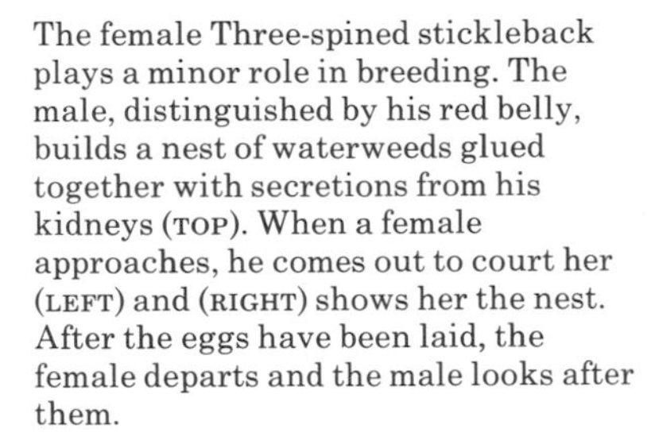

Courtship in the Three-spined stickleback. The male performs a zig-zag dance (1) to attract the female (2); he then leads her to his nest (3). The male points to the nest (4) which the female enters (5) and lays her eggs whilst being prodded in the tail by the male (6).

of a stickleback held in the water if its belly is not painted red. On the other hand, it will attack a crudely shaped model that does have red on the underside. The red belly is, for a stickleback, the equivalent of waving a red rag at a bull. Niko Tinbergen, who analyzed stickleback behaviour, recounts how his sticklebacks even displayed aggressively to a red mail van that passed the laboratory windows. However, a stickleback's reaction to a red-bellied intruder is much stronger if the intruder adopts the threatening posture of doing a headstand with its spines raised, and a fight may ensue.

The female stickleback in breeding condition presents a very different picture. She lacks the threatening red and her belly is swollen with eggs. The male's reaction to her is as definite as it is different from his behaviour towards another male. The females swim through the male's territory in small schools and, as they enter each territory, the male comes out to receive them. He swims about in front of them, leaping through the water and showing off his colours. Most move on but, if he is lucky, one female will stay. Then begins the courtship sequence, of each sex replying to the other's gestures. First the male performs a zigzag dance in front of the female. She comes towards him and adopts an upright posture that signifies her interest. The male now turns around and guides the female to the nest. She follows and he indicates where she should go by turning on his side with his snout in the entrance. The female enters and lies there with her head protruding from one side of the nest and her tail from the other. To encourage her to lay, the male prods her tail with his snout. After laying, the female departs and the male enters the nest to shed his milt over the eggs to fertilize them. Thereafter, he keeps watch over the eggs, fanning them with his fins and, when the fry have hatched, he guards them also.

The courtship of sticklebacks is a rigid ceremony and it has several functions. It identifies male and female, ensures that they are of the same species, that both are ripe for breeding and ensures that eggs and sperms are deposited in the same place at the same time. By spawning in the confines of the nest, sticklebacks are achieving the same end as

The freshwater mussel is a living incubator for bitterling eggs. The female bitterling lays her eggs in the body of the mussel by means of the long ovipositor and the male's milt is swept in after them on the current of water circulating through the mussel. In return, the bitterlings help to disperse the mussel's larvae.

LEFT: Oral insemination is the strange habit of this African mouthbrooder *Haplochromis burtoni*. It is one of many tropical fishes in which one parent carries the eggs in its mouth. The female of this species gathers her eggs in her mouth as they are shed. Meanwhile, the male swims nearby and sheds his milt. The female mistakes the bright spots on his anal fin for more of her eggs and attempts to snap them up. In doing so, she takes the milt into her mouth and the eggs are fertilized.

internal fertilization in that there is provision for a particular male's sperms, and his only, to fertilize the eggs of a particular female.

Strange happenings. Two other fishes are worth mentioning as examples of the strange ways that exist for ensuring fertilization of a female's eggs by one particular male even when fertilization is not internal. *Haplochromis burtoni*, one of the African cichlids, broods its eggs in its mouth. This is a widespread habit among cichlids, some of which are nicknamed mouthbrooders, as a consequence. In most mouthbrooders, fertilization takes place in the normal way by the male folding his body over the female to bring their genital apertures together. Then he or she quickly snaps up the fertilized eggs as they leave the female's body and before they can sink to the bottom. *Haplochromis burtoni* has developed a novel method of ensuring fertilization. The female alone collects the eggs as they are laid and before they are fertilized. However, the male is nearby waiting to spawn. He has a row of red egg-sized spots on his anal fin, which to the female looks like more of her eggs. She attempts to engulf them. In this she is naturally unsuccessful, but the result is that, as she opens her mouth to engulf these 'eggs', she takes sperm into her mouth where it fertilizes the eggs.

The spawning of the bitterling is a remarkable example of close co-operation, not so much between male and female as between two very different species. The bitterling is a small fish of European streams whose German name refers to its bitter tasting flesh. It is dependent for breeding on a freshwater mussel, known in Britain as the Painter's mussel. The male becomes brightly coloured in the breeding season and, at the same time, the female develops a two-inch-long, tubular ovipositor. Together, the pair search for a mussel and the female takes up position over its siphon. This is the tube through which the mussel draws in water from which it extracts food and oxygen. The female bitterling plunges her ovipositor into the siphon and lays her eggs in the mussel's body cavity. Meanwhile the male discharges his milt which is swept into the mussel on the ingoing current of water. The eggs are fertilized and the young bitterlings spend a month inside their living incubator before they swim out.

The bitterlings require the co-operation of the mussel because the female runs the risk of having her ovipositor gripped in the mussel's shell. She escapes this fate by repeatedly nudging the mussel and conditioning it not to react to the slight touch of the ovipositor. The relationship is far from being one-sided because, while the bitterling is laying her eggs, the mussel discharges its larvae. Each one, known as a glochidium, has a minute shell like a pair of jaws which clamp onto the bitterling. The fish's skin grows over the larvae and they are nourished on its juices until ready to drop off as a perfectly formed mussel.

Cuckoldry in a fish. Before leaving the breeding of fishes, it will be of interest to note the unusual spawning habits of the Atlantic salmon, in which the female has something of the proverbial 'two strings to her bow'.

Young salmon swim out to sea after spending some years in fresh water. They eat enormously and put on weight until they return as adults, swimming back upstream to the exact place where they were born. During the course of the summer the fat flesh, that makes a fresh-run salmon such a delicacy, feeds the ripening sexual organs. The silvery skin turns to pink with black flecks and the jaws of the males become hooked. The spawn-

Kissing gourami showing how they get their name. Kissing takes place between two males, or a male and a female, and is a demonstration of threat used in upholding territorial rights or in courtship.

The uphill struggle of the salmon. After feeding in the sea, salmon migrate upriver to the stream where they were born. There they spawn and often, die.

ing process is initiated by the female who digs a nest, called a redd, in the gravel of the stream bed. By violently lashing her tail she shifts gravel downstream and carves a shallow pit which she tests with her ventral fins. Now the male moves alongside and the female gapes while the male spreads his fins. This is the signal for spawning. She lays 800–900 eggs for every pound of her body weight and the male squirts out his milt. Fertilization must be extremely rapid because the sperms survive only a minute or so and will, in any case, be swept away by the current. Yet the male may not be the one to fertilize the eggs. Contrary to the general rule of a male ensuring that his sperms fertilize the eggs of his courtship partner, there is the occasional case of cuckoldry in which a male parr, a young salmon that has not yet been to sea, sneaks in and fertilizes the eggs. Because he is so much smaller than the adults, the parr can get nearer the eggs as they leave the female's body. This is a sort of fail-safe mechanism in case the adult male's milt gets swept away.

Frogs and toads. As the days lengthen and the spring weather gets warmer, amphibians leave their winter retreats and make for the ponds, rivers and ditches where they will spawn. Some species spawn at the nearest convenient stretch of water while others prefer traditional breeding sites. Common toads migrate in droves, following the same path year after year. The movement takes place at night but the route, where it crosses a road, becomes marked with squashed toads. The males move first and they take up pos-

Ménage à trois in frogs. When Common frogs gather in a pond to spawn, males fight to grapple with a female, clasping her from behind, so that fertilization can take place when the spawn is extruded. The competition for females may be intense and more than one male may seize a single female.

ition in the water in readiness for the arrival of the larger females who are swollen with developing eggs.

The aim of each male is to clasp a female in amplexus, that is to climb onto her back and grasp her securely with his forelegs. In the common European frog and Common toad mating seems to be on a first-come, first-served basis. The males attempt to grapple with any likely object, including fishes, and if one male grasps another male he is pushed away and a special grunt announces the mistake. The first female toads to arrive are often already carrying a male, to add to their laborious trek, and each is soon surrounded by a tight ball of males who may even drown her in their anxiety to fertilize her eggs. Eventually, however, order is restored and each female is properly paired off with one male. The male is virtually assured of fertilizing her eggs even though fertilization is external because both flex their backs to bring their cloacas together and the sperms mix with the eggs as they are extruded.

In some parts of the world the breeding season of frogs and toads is immediately recognizable by the incessant chorus of croaking, which can become extremely nerve-wracking in its monotony. The main function of the croaking is to act as a beacon, telling the amphibians of one another's position. Nevertheless, some species have a varied repertoire of calls. The bullfrog of North America has seven calls, each used on different occasions.

The familiar croak for which the bullfrog is named, a booming sound audible over a distance of half a mile, is the mating call that makes up the chorus of the male bullfrogs. It is best described by saying that it can be imitated by shouting 'rum' in a deep voice into an empty barrel and it is produced by a vigorous exhalation forcing air across the vocal cords and into the two vocal sacs under the chin. The sacs momentarily bulge, then deflate as the air is returned to the lungs for the next croak.

The bullfrogs' chorus of mating calls can be heard each evening during the breeding season as the frogs come out of their daytime hiding places along the banks and take up stations along the shore. The male frogs adopt a characteristic high-floating pose caused by the inflation of the lungs for croaking and which shows off the yellow chin. The females are attracted by these calls, preferring a chorusing group to a single male. When she finds a group of males, a female swims among them with body submerged and only her head showing, in contrast to the high floating males. Eventually she swims to one particular male and allows herself to be clasped.

How the female bullfrog decides which male to approach is not known. The territorial behaviour of the males ensures that she has a choice and can mate without the difficulties that beset female Common toads. The only indication of how choice is exercised comes from some recent experiments on Tree frogs. Male Pacific tree frogs gather in groups, with 15–20 inches between each frog. They call in bouts, croaking in unison for a few minutes then falling silent. Female frogs are attracted by the chorus and approach one particular male and allow him to clasp them. This male is particularly attractive because he is the most vigorous croaker. He is the 'bout leader' who initiates each bout of croaking and is the last to stop at the end of the bout. He also croaks faster and louder than his fellows. The $64,000 question is: if vigorous calling is an indication of a male's superiority, how are the two linked? Presumably the 'bout leader' is dominant to the others and, somehow, characters that make him a suitable father to the next generation

The male newt attracts the attention of the female by fanning his tail so that secretions from glands in his skin are wafted to her. If the female is responsive she follows him and eventually he deposits a spermatophore which she picks up.

ensure his dominance which is indicated by his croaking prowess. The link between fitness to be a parent and social dominance is an intriguing problem and one that is basic to our story of courtship. As yet it is largely unanswered except in cases where the obviously strongest male wins contests that prevent other males mating.

The delicate display of newts. The newts are tailed amphibians familiar to anyone who has foraged with a net in ponds and ditches. During the summer months they share these quiet watery habitats with their relatives the frogs and toads and, like them, they usually stay in the water only as long as is necessary for reproduction. The rest of the time is spent on land, usually hidden in low growing vegetation or in crannies and crevices. Newts can be kept in aquaria through the breeding season and watching their courtship and egg-laying makes a delightful indoor pastime. Their life history is very similar to that of frogs and toads but there are striking differences in the rituals of courtship and egg-laying. There is no scrummaging around the female, with jostling and kicking for the privilege of amplexus. Newts are lithe swimmers, compared with the clumsy breast-stroking toads and frogs, and their courtship is a dance between a pair of animals that completely lacks the aggression seen in other animals.

A basic difference in mating of newts and frogs and toads is that newts practise internal fertilization. The male lacks a penis but, in a manner reminiscent of scorpions, he places a spermatophore on the floor of the pond and induces the female to pick it up in her cloaca. The object of the courtship is, then, to stimulate the female and entice her into position over the spermatophore.

Courtship is prolonged, the female being unresponsive when the male approaches and sniffs her body. She tends to swim away from him and he has to follow and attempt to take up position across her path. After a while, the female stands quietly and the male can start to woo her with three distinct displays. The first is visual and consists of the male poising in front of her with his tail, a showy appendage decorated with a frilled crest and broad spots, held at an angle. This is followed by a

RIGHT: Amphibians such as this frog spend much of their lives on land but must return to water to breed. The males gather near water and attract females by loud croaking. Air is pumped between distended vocal sacs and the lungs via the vocal cords.

A few frogs carry their tadpoles on their backs.

violent whip of the tail which create a surge of water so powerful that it may knock the female backwards. In the most obvious display, the tail is folded alongside the body and is rapidly fanned to create a stream of water flowing towards the female's snout. The cloaca is opened and secretions from special glands are wafted to the female.

The displays are continued for some time until a change in the female's behaviour marks the start of the next phase of courtship. She begins to approach the male and it becomes his turn to move away, but he continues to display as he swims backwards. Finally, the male turns and creeps forwards, with the female following close on his heels. This is the final phase of courtship and is a delicate operation in which the male newt deposits his spermatophore and positions the female so that she can pick it up. He proceeds slowly for a few inches then stops and quivers his tail. The female advances until her snout touches the male's tail and, in response to this signal, he extrudes the spermatophore. From this position the female has to manoeuvre until her cloaca is over the spermatophore. To accomplish this, the male advances and turns sideways so that he has moved exactly one body-length forwards. The female advances until she is just touching him and her cloaca will be automatically positioned over the spermatophore, which she picks up.

Independence from water. The egg and tadpole stage of the amphibians ties them to a watery environment. Outside the breeding season, adults of most species lead a com-

Not all amphibians breed in water. This African Tree frog lays its eggs on a leaf. They are protected by a thick liquid which is whipped into a foam by the frog's back legs. When the tadpoles hatch, they fall into the pool beneath.

Frogs with their spawn. After a few days of frenzied courtship and spawning, the water is deserted by the adults and left to the developing eggs.

pletely terrestrial life. Many, it is true, prefer a moist environment but there are toads that survive in deserts by burrowing and storing water in their bladders. There are a number of amphibians that are experimenting, as it were, with methods of dry land reproduction. The Kloof frog of South Africa attaches its eggs to a vertical rock face overhanging a stream and the emerging tadpoles fall into the water. The male Midwife toad carries the eggs wrapped around his legs and returns to a pond to moisten them at intervals. Other frogs have become completely independent of water for breeding. Rattray's frog lays its eggs in a nest and the tadpoles develop in the jelly that originally surrounded the eggs, while certain frogs of the genus *Platymantis* live permanently in trees of the Philippine forests and omit the tadpole stage. The eggs are laid in leaves and fully developed froglets hatch out after several weeks.

Omitting the larval, tadpole, stage is essential for the full development of land-living and was an important stage in the successful evolution of the reptiles from one branch of the amphibian stock. At the same time, internal fertilization was necessary for the development of an egg which could receive an impervious leathery or chalky shell after fertilization. The drought-resistant egg, together with waterproofing of the adult animal, was crucial to the explosion of reptile forms that made them the dominant land animals for a period of 100 million years. Since then they have been replaced by their descendants, the mammals and birds, which also depend on internal fertilization and waterproofing. The vertebrates can, therefore, be divided into the fishes and amphibians on the one hand and reptiles, birds and mammals on the other. The former lay eggs protected by a gelatinous envelope, and containing little yolk because they soon hatch into larvae. The majority of the latter, however, lay

The ponderous mating of the Leopard tortoise. He has to make sure he does not slip off the female's shell.

an egg with a shell and plenty of yolk to sustain the developing embryo until it hatches beyond the larval stage.

Once the reptiles developed dry-land reproduction, there was no turning back and even the turtles which spend their lives at sea must laboriously heave their heavy bodies up a beach to lay their eggs above the high tide mark. Courtship among the land tortoises is necessarily a ponderous, graceless affair and is extremely simple. The male walks up to the female, who is often considerably larger than himself, and butts her with his head. Some strong stimulus like this must be needed to penetrate to the female's sensibility and a considerable amount of butting is needed to make her receptive. For further stimulation, the male tortoise also bites her legs. The act of mating necessitates a balancing act to prevent the male sliding off the dome of the female's shell. To aid this, the rear part of the male's plastron, the underside of his shell, is concave to fit to some extent the female's carapace. Together with a thickened tail, this is the only outward physical difference between the sexes. The male Box tortoise rears up vertically behind his mate and hooks onto her shell with his long hind claws while she assists by pressing her legs against his ankles.

Aquatic chelonians, as terrapins, water tortoises and marine turtles are jointly called, have more scope for courtship as their buoyancy in water gives them greater freedom of movement. The Red-eared terrapin swims backwards in front of the female and tickles her chin with his long claws. When she has been aroused she sinks to the bottom and the male descends to her.

By comparison with the ponderous chelonians, the slender, agile lizards have a courtship full of colour and movement. Anoles are a group of exclusively American lizards of the iguana family. The Green anole is often called the 'American chameleon' because of its ability to change colour. Like the true chameleons, the anoles alter colour according to temperature, background colour and emotional state. During the breeding season males assume a courtship dress. Each stakes out a territory which he defends against other males by inflating a throat pouch so huge that he has to throw back his head to prevent it scraping the ground. The throat pouch is coloured blue, green and yellow and is shown to full effect by the anole standing on tiptoe parallel to his rival, with legs stretched and tail wagging.

A female anole is treated to the same display but her behaviour is the opposite to that of a rival male. Like the female cichlid fish, she advances coyly, head held to one side and with no sign of a threatening throat pouch. The male approaches from the rear, grabs her neck in his jaws, puts one foot over her body and twists his tail under hers to bring the reproductive organs together.

The African counterparts of the anoles are the agamas which also become quarrelsome

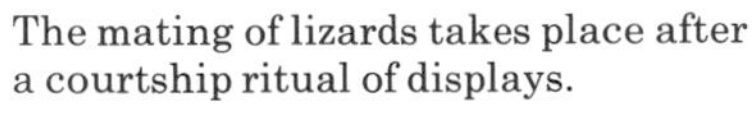

The mating of lizards takes place after a courtship ritual of displays.

The Blue-throated tree agama displays a blue throat pouch and nods his head to threaten rivals and attract females.

Two African boomslangs, tree-living snakes, coil around each other. The male is attracted by the odour of the female.

in the breeding season. They display to each other with comical head-bobbing as if they were intent on keeping fit with press-up exercises. Agamas are polygamous and each male keeps a harem of half a dozen females in his territory. Much of his time is spent in defending this territory both by display and by exchanging blows of the tail, sometimes of such violence that jaws and tails may be broken. However, each evening a truce is called. The courtship colours are replaced by a dull brown and all the males retire to a communal roosting tree where they spend the night amicably.

The courtship of snakes lacks the colour and movement of the lizards. They rely more on the sense of smell and follow scent trails laid by other snakes. Some male snakes mate successfully if they are blindfolded, but still able to smell the female. Smell is not sufficient for other species and identification of the female and stimulating her to breed is carried out by touch. The Red-bellied snake of North America indulges in a form of massage. He runs his chin up the female's back, while flicking his tongue in and out. When his head is resting on her neck, he throws a loop of his body over hers and twists his tail under hers to bring the reproductive organs together. Then a series of waves run up his body becoming more violent so that he is moving against her body at many points.

The mating of snakes is particularly noteworthy because of the copulatory play. It was discovered only in 1974 and has been found so far in three American snakes – two Garter snakes and a Water snake. After copulation, the male inserts a plug made from kidney secretions into the cloaca of the female. It acts as a kind of internal chastity belt in preventing further copulation. As these snakes gather in mating swarms of mixed sexes, the plug ensures that a female is fertilized by the first male to mate with her.

Birdsong is a delight to human ears but its meaning is menacing. This male European blackbird is singing to show its ownership of a territory. Other males must stay away. However, the song also identifies it to female blackbirds and may stimulate them to come into breeding conditions.

SONG AND DANCE ON WINGS

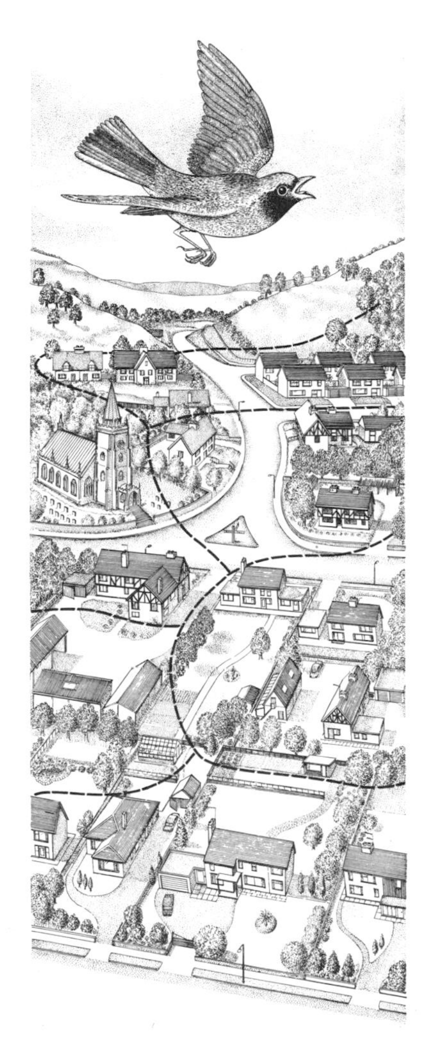

The territories of songbirds are large enough to provide food and resting positions for the entire family and are vigorously defended.

As we go up the animal scale, reproductive behaviour becomes more and more complex. The nervous system becomes more refined and the sense organs become more finely discriminating. But what places the courtship of birds at the peak of sophistication and makes it such a joy to the human eavesdropper is the development of the voice and the variety of form and colour achieved by the use of feathers. Voice and appearance are used in many different ways to enhance the threatening and wooing messages of male birds, and, luckily for us, both are so arranged as to give pleasure to human eyes and ears. As in the vocal and visual 'languages' of other animals – the chirping of crickets, croaking of frogs and 'beckoning' of Fiddler crabs, for instance – the songs and displays of birds have a manifold purpose. They allow one animal to make its presence felt to another, they threaten and repel rivals and they attract prospective mates and synchronize their breeding behaviour. In short, they are used for defending territories and in courtship.

Territorial behaviour was discovered and studied first over 50 years ago, by the amateur ornithologist Eliot Howard. Yet there is still disagreement about its function in the lives of animals. Evidence is conflicting but it is quite possible that, as so often happens, it will be found that there are several functions used in different situations. For one thing, there are several different kinds of territory. The large territories of familiar songbirds, for instance, provide most of a bird's needs. They are areas where the birds rear their families and obtain their food. One very useful function of such a territory is that the resident birds become familiar with the terrain and they know where to find food and can instantly flee to refuges when danger threatens. A defended territory also enables a pair to rear a family without interference. At the other end of the scale, there are small, short-lived territories which are used solely for courtship and mating, with nesting taking place elsewhere.

The main controversy over the function of territories is whether they are useful in controlling the size of the breeding population. For instance, is the available habitat divided up so that every pair gets a territory of some size and all can breed? Or are the territory sizes fixed so that the area of habitat is parcelled out among a limited number of pairs and the surplus animals are forced to breed in less favourable habitats or not breed at all? The argument cannot be resolved until we learn more of the habits of many kinds of birds but it is not of importance here as we are more concerned with the courtship behaviour that takes place within the territory once it has been established. It is sufficient to say that it is part of the male bird's reproductive strategy to set up a territory where he can have exclusive access to one or more females.

The courtship and mating of birds fall into two general categories. In some birds, male and female pair up and remain faithful for the entire breeding season and both help to rear the family. Others have one-parent families. After courtship and coition, the pair splits up and one bird, usually the female, rears the family alone.

Food supply seems to be the factor determining whether a bird shares parental responsibilities and this has a significant effect on the form of courtship.

Birdsong. Birdsong is not easy to define. To most of us it is the pleasing patterns of loud, rhythmic sounds that herald the new day throughout the spring and summer months. We call it song because, by its nature, it has characteristics similar to music, but such distinction leaves out much of what an ornithologist would call song. The monotonous

call of the cuckoo, the cawing of crows and the harsh screams of gulls are, strictly speaking, birdsong. They fulfil the same functions as the melodious strains from robins, blackbirds and the many warblers. They are a means of advertisement, communicating the position of the bird and indicating its ownership of an area and its need for a mate. For this reason, singing is most persistent in the early part of the year when the unmated males are jostling to carve up the land into territories. By the time eggs are being laid, everyone has settled down and only an occasional reminder is needed to keep the territory boundaries intact. As a general rule, we can say that birdsong is well developed among woodland birds, particularly those species that skulk in the undergrowth, like the wren. They have to rely on sound for their neighbours to locate their position. Unfortunately we can immediately think of birds that make themselves conspicuous while they sing. There is the Song thrush that sings from the top of a tree and the pipits and larks that fly up in a song--flight.

The greatest development of song is in the true songbirds, the suborder Oscines of the order of perching birds, the Passeriformes. The remarkable vocal ability of these birds lies in the structure of the syrinx, the equivalent of the mammalian voice box or larynx. Sounds made by birds originate entirely in the syrinx and are not modified in any way by the tongue or mouth, as in human speech. So fine is the control of the syrinx that some birds can sing more than two notes at once. The Gouldian finch, a weaver-finch often kept as a cage-bird, matches its strikingly brilliant plumage with a three-fold song. It sings two simultaneous tunes, backed by a bagpipe-like drone. There is some accessory modification of sounds, however. The lengthened windpipe of Trumpeter swans and Whooping cranes are responsible for their deep calls and the Sage grouse uses inflated throat pouches to act as resonators for its loud 'popping' calls. 'Song' in the strict sense need not be vocal. Some birds use instrumental music. The Greater spotted woodpecker drums on trees with its bill, and the American woodcock and European snipe whistle and bleat in flight by allowing special feathers to vibrate in the slipstream as they dive.

Like other forms of display, birdsong does not act only as an advertisement to either sex. It is a stimulant. The season of song starts as the testes become active. Outside the breeding season they lie shrunk and inactive in the bird's body. Then, as days lengthen, they undergo a phenomenal increase in size and

Gannets' nests are tightly packed in cliff top colonies. There is bitter rivalry between the pairs and nests are spaced so that the owners are just out of pecking distance.

ABOVE: Male ruffs gather to display on 'hills'. Each has a small territory where it shows off its fine plumage. The females or reeves arrive later and choose a mate from the selection of gaudy males. Certain males are dominant and are preferred by the reeves, who queue for their favours.

RIGHT: A male Superb lyrebird displays on a mound of earth, showing off his fan of tail feathers. At the same time he calls out a continuous song which includes the mimicking of other birds.

LEFT: Among the African weaverbirds, nest-building is a prelude to courtship. After building his nest of fine strips of leaves, this Black-headed weaver will hang upside-down from it and display to passing females.

LEFT: One of the best known of all bird displays. Peacocks, the males of the peafowl, have been kept for centuries for the brilliance of the train of feathers that unfolds during courtship.

After a spectacular mass courtship involving thousands of pink-tinged birds, flamingos pair off before laying a single egg.

they pump out sex hormones to initiate the many activities of the breeding season. Singing, in turn, stimulates the female to achieve breeding condition, probably acting on the ovaries via the brain and the hormone-secreting pituitary gland lying on its underside. The female canary needs to hear her mate's song for her ovaries to develop. She cannot lay unless she has a nest hole which is dark inside – a nest box is adequate – but eggs will still not form if she cannot hear the male warbling. The more he warbles, the quicker the ovaries ripen and the eggs develop.

Later in the season, after the eggs have been laid, song serves to maintain the bond between the pair in those birds where the male stays to help rear the family. The exchange is sometimes two-way because, although most female birds are silent, a few are songsters and perform duets with their mates. The Eastern whipbird of eastern Australia is named after a song that starts as a pure note increasing in volume and ending suddenly with a sharp crack. The female then replies with her own notes. The duet helps to keep the pair together in the short term, as they forage in dense undergrowth, and in the long term by preserving the pair-bond.

Displays of finery. Every function of birdsong is paralleled by visual displays. As birdsong is a delight to our ears, so the displays of many birds are a pleasure to watch. Graceful movements of neck and wings, perhaps even dance steps, are augmented by elaborate plumages or patches of colour designed to catch the attention of other birds. Both the postures adopted in display and the form of plumes and colour are

Mutual preening, or allopreening, is a common habit among pairs of birds, as shown by this pair of Black-browed albatrosses.

important for species recognition. A male Black stork was once seen to court a female White stork in the unnatural circumstances of a zoo but, even with the incentive of being confined in the same cage to stimulate cross-breeding, the union was never consummated. The Black stork courted with sinuous movements of the neck and the White stork responded by laying her neck over her back and rattling her bill. Neither display was recognized by the other species, so nothing further happened. The importance of signals in demonstrating the sex of a bird is further shown by artificially changing plumage patterns. The male Yellow-shafted flicker, an American woodpecker, wears a black 'moustache', a pair of black marks on the throat. If 'moustaches' are experimentally glued onto a female flicker, even one that has already been paired, her mate will treat her as a male and drive her out of the territory. When she is caught again and the moustaches removed he will once again court her.

Postures, plumes and colours have proved a very fertile basis for sexual selection. If these attract females, there will be an evolutionary pressure to develop these in more spectacular form and, as a result, we see such extravagances as the showy train of the peacock and the 'runaway' evolution of the Birds of paradise. When Bird of paradise plumes started to appear on the European millinery market, ornithologists could not believe they were real. It did not seem possible that such ornate structures could be natural or could have any function. In fact, we now know that many Birds of paradise display communally and the female visits the display grounds to select a male as mate. In this situation sexual selection can be most effective and, although the ornate feathes of the males must be a hindrance to flight, the disadvantage is outweighed by their value in obtaining a mate.

Many birds have a repertoire of displays, each appropriate to a different situation. A particular display is used to indicate a mood or intention and the bird at which it is displaying can react as appropriate. The meanings of displays and their underlying causes have been the subject of considerable ornithological research, using the gull family in particular.

Birds stay apart from each other outside the breeding season. Even when living in flocks they do not come too close. Then breeding behaviour forces males farther apart as their territorial instinct asserts itself. On the other hand they have to allow females near them, even into intimate contact if breeding is to take place. The displays between the birds – between male and male or male and female – are designed to expedite repulsion or attraction. In both situations a bird is motivated by two urges or drives. He has an aggressive drive to chase the other bird but he is also afraid that the other may attack back. The result is an intermediate situation: he threatens but does not attack and the mixture of the two drives is shown in his displays.

A male Herring gull seeing an intruder entering his territory walks towards it in an 'Upright posture'. His wings are held slightly

Courtship starts with each side wary of the other. They threaten one another not to come too close but hostility gradually declines. These Black-headed gulls are threatening with forward and upright postures – later they will turn their heads away to hide the threatening black masks.

outwards, an indication that he is ready to take off, and his bill is tilted down, ready to peck at an adversary, but he is not willing to go the whole way and attack. The message is clear to the intruder who may threaten back or, if he is not anxious to get involved, retreat with neck withdrawn and head held low in submission. When a gull is very aggressive but is still not willing to attack he tears at the grass around him.

The reaction of a male gull to a female entering his territory is the same as to a male. He sees her as a potential danger and threatens her but she turns his aggression aside by assuming a submissive posture. Over a period of days, the two birds get gradually used to each other, threatening displays subside and mating can proceed.

On living together. Female birds are placed under a great strain during the period in which their eggs are forming and being laid. They must spend time in courting and nest building yet still forage to find sufficient food for their bodily needs and for egg formation.

LEFT: The preliminary to courtship in the Wandering albatross is the spectacular sky-pointing display. The pair face each other with the 10-foot span of wings arched and bills pointed skywards.

BELOW: A male Wandering albatross displays to a prospective mate who hides her uncertainty in preening. A second female approaches. Young albatrosses spend several years in promiscuous courting before choosing a mate but, once formed, the bond is permanent.

Male African dabchick feeds his mate. Courtship feeding helps to cement the bond between two birds and the extra food helps the female to build up the eggs.

The strain can be relieved to a greater or lesser extent by the male feeding the female during courtship. This courtship feeding probably has a certain ritual value in maintaining the pair-bond but it undoubtedly has a more practical value also.

Among terns, it is the custom for the male to carry a small fish in his bill and display it to prospective mates. Later in the season, when he has 'married' one of them, she will become increasingly dependent on food brought by her mate, particularly when involved in laying the three eggs. The more food he brings, the larger will be the eggs she lays, and, consequently, the larger will the chicks be. As large chicks are more likely to survive, it is an advantage to the female tern to choose a male who is a good provider. She makes the choice on the basis of the male's performance during the fish presentation display.

Courtship feeding occurs only when male and female stay together for some length of time. The male is investing more in his mating than the supply of genetic material and, for the first time, we are seeing the development of a long-term bond between the sexes, the aim of which is to increase the chances of survival of the offspring. Courtship feeding is only one of several strategies now open to the male bird to further this end. He can also help incubate the eggs, feed the young and protect the nest from danger. In fact, in such a monogamous situation, he is investing about as much time and energy in the production of offspring as the female and both are benefitting, in terms of reproductive strategy, by the care they give to their young.

The continuing trend of reducing the number of ova and increasing the care given to offspring takes a great step forward in birds and reaches its ultimate in the great albatrosses which lay a single egg every other year. The egg of the Wandering albatross, the largest of them, takes 11 weeks to hatch and the chick spends a further 10 months in the nest before it fledges. By the time the parents are released from the chores of raising one chick, there is not sufficient time for them to recover from the strain of feeding the chick and prepare for the next egg-laying season. So they miss a year and concentrate on getting into good condition for the season after.

As both sexes of the Wandering albatross invest a good deal of time and energy in raising the single chick, it is an advantage to both to ensure that the partner is going to honour the contract and see the business of rearing the chick through to the bitter end. The method of achieving a strong and lasting bond is to take a long time over courtship. Then, if either bird is found wanting, its prospective mate can drop out without having wasted his or her investment. So we find that courtship also is a lengthy process in Wandering albatrosses.

Young females of this species come back to

For the breeding season, the male Jackson's widow-bird grows an unwieldy tail streamer. To attract females, he flies about with the tail waving like a banner.

the breeding grounds on windswept Sub-antarctic islands when four or so years old but they will not lay their first egg until at least twice this age. Each young male finds a suitable place to start the foundations of a nest. He builds a pile of moss, grass and soil and, standing on it, proclaims his availability with a spectacular 'sky-pointing display'. He points his bill at the sky and, holding his 11-foot span of wing outstretched, utters a clear, far-carrying whistle. Meanwhile, female albatrosses are circling overhead and swooping low over the conspicuous male and, eventually, one decides to take a closer look, lands and walks up to him and allows herself to be courted. At this stage, courtship does not proceed far and a female will walk from male to male, spending more and more time with one particular male as the days go by. Courtship proceeds hesitantly at first because, as with the Herring gull, the male has an aggressive streak which has to be placated. His 'sky-pointing' may change into a vicious lunge at the female, whose aim is to draw close and appease him. In this, she stretches out her neck and attempts to nibble his neck feathers. At first she is on guard against a peck from his six-inch, hooked bill but, at length, both birds are sitting side by side, gently nibbling each other's necks and softly

A pair of Arctic terns display where the nest bowl will eventually be excavated.

After driving off a rival, the male Black-necked swan returns to his mate and lifts his chin in a 'triumph ceremony'.

The brightly-coloured Red-breasted merganser impresses his mate with the 'curtsey display', depressing his breast in the water and spinning like a top. The pair splits after mating and the drably-coloured duck incubates the eggs by herself.

calling. Even now, the pair may split up but there comes a time, one summer, when the female will finish the lining of the nest and, a little later, lay a single egg.

Once the union has been consummated, the albatross pair will not split up until the death of one partner. Their efforts at breeding benefit greatly from their combined experience, particularly as their familiarity with each other cuts down the time needed for each renewed courtship. A general experience of breeding also helps; older birds are more successful at rearing their chicks. They carry out their incubation duties more meticulously and probably bring more food to the growing chicks.

A permanent, season-to-season pair-bond exists in many seabirds, and long-term researches on kittiwakes in eastern England show that the longer a pair stays together the more successful it is at breeding. Well established pairs arrive at the colony earlier in the season and lay eggs earlier, so getting a head start. Such pairs also raise more chicks.

As more seabirds and other long-lived birds are studied intensively, by means of fitting them with combinations of coloured rings to allow easy identification of individuals, so permanent pair-bonding or marriage is being found to be more widespread than was formerly supposed. Pairs of penguins, albatrosses and kittiwakes move their nests towards the centre of the colony as they get older. The younger, unstable pairs nest on the fringes of the colony where there is bickering and disturbance as the young, inexperienced birds settle down to the responsibilities of parenthood, while the older, more experienced birds in the centre can get down to the serious business of nesting without delay and in peace.

The jackdaw's long fidelity has been charmingly chronicled by Konrad Lorenz in *King Solomon's Ring.* There is a rigid hierarchy or 'peck order' among jackdaws in which lower-order members of the flock must defer to their seniors. When two jackdaws mate, they share everything including social status, so that an inferior female gains rank when she 'marries' a superior male. However, it is a law of jackdaws that no male can ever mate with a female who is his superior. Among the common songbirds of town and country, permanent pairing is not the rule. These birds do not live very long so there is less chance of both members of the pair surviving from one year to another. If they do pair again it is only because they tend to return to the same place where they nested before and are therefore likely to meet again.

Sharing mates. Monogamy, the mating to a single member of the opposite sex and the assurance of sole access to that individual for breeding purposes, is a particular feature of the social life of birds, more so than in the mammals. Ninety per cent of bird species are monogamous, these being species in which both parents need to feed the young, Polygamy, which is far less common, may involve one male mating with two or more females, when it is called polygyny, or two or more males mating with one female. Polyandry, as this last is called, is more rare than polygyny. Polygamy occurs where, for some reason, food may be so abundant that one parent can cope with feeding the brood or it may be that the chicks are able to feed themselves on hatching, as in chickens, ducks and pheasants.

Even among traditionally monogamous birds, polygamy sometimes creeps in. There are rare instances of male Brown skuas taking two mates who lay clutches of eggs in the same nest, and a Reed bunting may sometimes have two mates either sharing a nest or

A pair of Red pied stilts in their aquatic environment. Stilts are wading birds that feed in both salt and fresh water and in flooded grasslands.

RIGHT: Whooper swans bathe as a routine after copulation.

BELOW: Mutual preening by spoonbills. The action helps to cement the bond between the birds.

The courtship of kingfishers consists of aerial chases, the two brilliant little birds circling and dashing to and fro. At the climax, the female perches and the male alights on her back, holding her neck firmly with his bill.

rearing young in separate nests. Sometimes he helps to feed both broods. This sort of polygamy occurs when food is particularly abundant. In the Red grouse of Scotland, most males are monogamous but some manage to establish larger than normal territories and then attract two females. There is also serial polygyny, as in Brewster's blackbird of North America, in which the male pairs with one female and guards her against the advances of other males until she has laid, then courts and guards a second female. When the chicks hatch, he helps to feed both broods. The Common starling is normally monogamous but at least one instance is fully documented in which one male had three 'wives' and managed to feed all three as well as help all in the incubation and feeding of the chicks.

Finally, in several of the families of birds, one sex is freed from parental duties and is needed for mating alone. Once this has happened, the redundant parent, usually the male, can ignore the restraints of parenthood and concentrate on winning as many partners as possible. And, if some males obtain the favours of two or more females, others must fail to attract any, assuming the sex ratio is equal. Therefore, there will be competition between the males for this privilege. The competition results in intense sexual selection with the males evolving ostentatious plumes, voices or displays to impress the females. That is the situation in Birds of

White plumes that once adorned society hats are part of the cock ostrich's display. At the same time his throat expands and colours pink. He will mate with several hens and help rear their young.

paradise. Competition has made courtship stratagems incredibly elaborate.

Polygyny has been studied intensively in the grouse family where, apart from the Red grouse and the Ruffed grouse, the males have renounced parental duties and concentrated on ornamentation and communal displays to ensure the maximum number of offspring through the dominant or 'fittest' males impressing as many females as possible. The males of these grouse gather at traditional display grounds called arenas, or leks, after the display grounds of the Black grouse.

Within the arena each male holds a small territory where it displays first to defend the boundaries against other males, then to attract females. The displays involve loud calls, elaborate plumes and colour. Prairie chickens inflate orange airsacs and raise the neck epaulettes and tail. They run to and fro and spin in circles with wings drooped, while the tail is raised and lowered with a loud click, but the main noise comes from a continuous booming which can be heard over a distance of four miles. Black grouse leks can be located from a distance by the continuous bubbling and wheezing calls of the assembled males. The male Black grouse, or blackcock, has a magnificent glossy black plumage with a lyre-shaped tail and red wattles over the eyes. In display he fans his tail, extends his wings and blows out the undertail coverts to make a conspicuous white 'powderpuff'.

Male grouse would seem to be extremely

vulnerable when they gather on leks and deliberately draw attention to themselves; but they are wary and observations over a number of years at Prairie chicken 'booming grounds' have shown that remarkably few are killed by predators. Observations also show that a minority of the males present take part in the majority of matings. At three leks of Sage grouse, 7 per cent of the males mated with over 80 per cent of the females and the majority never got a chance to copulate. The natural question is what makes these few choice birds so successful with the females? They must have some advantage over other males and the females must recognize this advantage. The nature of the advantage and its recognition has long been a problem for those seeking to understand the workings of sexual selection but studies by two Dutch ornithologists have shed light on the matter.

The male Black grouse flock parcel out the lek unevenly so that some get no territory and others can find a territory only well away from the others. On the main lek, there is a gradation from large territories on the periphery to small, closely packed territories in the centre. As a male gets older and survives more breeding seasons, his territory moves nearer the centre. So the central males are the wily, experienced birds which have shown themselves to be natural survivors who can cope with the dangers and problems of life. The females, or greyhens, fly straight into the centre of the lek where males are densest and spectacular displays and fights are frequent, so they are immediately selecting the cream of the blackcocks. But this is not the end of the story because, of the central cocks, the most experienced have the best courtship tactics. They spend more time displaying when there is a greyhen present and have a better technique for wooing her.

As we have seen with other birds, there is a natural reluctance for two individuals to come together. The female is attracted but is also afraid to come too close. So the successful male is he who corrals the greyhen on his territory and reacts sensitively to her behaviour. He tries to get close to her but, as soon as she starts to retreat, he withdraws to put her at ease again. Thus the blackcock that become parents are those that are cleverer in all departments in life and are clearly most fitted to propagate the species.

Among the many polygamous species of birds, there are a few oddities where the female courts the male. The phalaropes are monogamous but the female phalarope is more brightly coloured than the male and leaves the father to rear the family. She takes the initiative in courtship and keeps other females away from the male. When, however, the male attempts to mount her, she goes 'coy' and leads him away. This is a device to ensure that he is free to look after her eggs. If he was already tied to a nest he would not be able to follow her.

The reversed sexual role probably arose from a situation where, after both sexes were involved in rearing the family, conditions changed so that one parent could be freed. In most birds faced with this situation, the male has opted out but in some instances the female freed herself and has opened the way to polyandry. In tinamous, partridge-like birds of Central and South America, the female goes from one male to another, leaving each with a clutch of eggs to tend. The female Pheasant-tailed jacana or lily-trotter is a long-tailed waterbird who has a harem of males, each smaller and more inconspicuous than herself. Polyandry is possible because the jacana's marshy habitat is extremely rich, so the female can readily generate several clutches of small eggs and the young can feed themselves on hatching.

At dusk on summer evenings, woodcock fly up and down woodland rides and clearings. Known as roding, these flights are a territorial display.

SEXUAL COMPETITION BY MAMMALS

Both birds and mammals are descendants of the reptiles but they have evolved along different lines. The brains of birds are largely given over to the precise control of flight and their behaviour is consequently simple and rigid. By contrast, mammals have developed a high degree of intelligence and have considerable powers of learning. Similarly, birds need good vision for navigation and steering during flight but mammals have, literally, kept their noses to the ground and they rely on the sense of smell. This explains why birds have evolved elaborate, eye-catching displays whereas the courtship of most mammals is more prosaic. Birds use visual signals but mammals communicate by pheromones which are beyond the power of our sense of smell to detect. Furthermore, mammals tend to be nocturnal or secretive, so we see little – in many instances, virtually nothing – of their private lives.

Mating in mammals often appears to be very casual. Preliminaries to courtship may seem sketchy, if not entirely lacking. The reasons are two-fold. We cannot appreciate the communication by pheromones that passes between the participants; and, of greater importance, the reproductive strategies of mammals are different from those of birds. Monogamy is rare and the male does not often help rear the ensuing litter. This is a result of the mammals' unique system of feeding the young. By definition, a female mammal feeds her young on milk secreted from her body and the male cannot therefore contribute to his offspring's nutrition except by feeding the female or providing food when the young are weaned. Consequently, the male invests very little time and energy in reproduction: so, unlike the Wandering albatross of the previous chapter, an elaborate courtship is not needed to ensure that the correct choice of mate is being made. The mammalian strategy typically follows the pattern of the Black grouse in having a hierarchy where a few dominant males mate with most of the females.

Cycle of acceptance. The female mammal has a breeding cycle, called the oestrus cycle, in which her reproductive organs and behaviour go through a pattern of change. For most of the cycle, she is in an-oestrus and is sexually inactive but, at one point during the cycle, she comes 'on heat' or 'in oestrus' and accepts the advances of the male. At this point her ova are ripe and ready for fertilization and the lining of the uterus is prepared for receiving the fertilized ova or zygotes. She shows her condition by her behaviour and usually by changes in the appearance of the genitalia.

The oestrus cycle stops when the animal becomes pregnant but, so long as she is not pregnant, the cycles follow in regular order, annually in most mammals but at intervals of a few weeks in rodents. There may be a specific breeding season when all females

The platypus, one of Australia's egg-laying mammals.

RIGHT: Two young Elephant seal bulls joust by banging chests. The impact makes their blubber wobble like jelly but no harm comes to the seals. At this stage the clashes are not serious. The seals are practising for the future when they will defend harems of cows.

ABOVE: Giraffes necking. As a trial of strength between males, two giraffes push against each other and only rarely come to blows.

come on heat, as in seals and badgers, but a few can breed all the year round, as in the giraffe, otters and dogs. Ovulation, the release of ripe ova into the reproductive tract, is not always automatic. Rabbits and ferrets ovulate only when mating has taken place.

A major part of the courtship behaviour of mammals consists of the male finding out whether the female is on heat and ready to accept him. Oestrus is usually signalled by pheromones secreted from the genital region. The male will regularly investigate the female with his nose and many of the hoofed mammals deliberately smell the urine of the female, to the same end. At the same time they make a characteristic grimace known as flehmen. This is a German word for which there is no English equivalent. It looks like an expression of disgust as the animal raises its head, rolls up its lips to expose the gums and wrinkles its nose. The significance of flehmen is not known.

The pheromones may attract males from a considerable distance as when a bitch comes on heat. The signal is first transmitted before she is properly on heat. The dog starts to pay attention but is repulsed by the bitch. He stays with her, following her about and occasionally attempting to mount her. So long as she is not receptive she growls when he does this and moves away but she does not reject him outright. Later she plays the coquette, almost inviting him to mate but without adopting the proper mating position. Even-

LEFT: The courtship of carnivorous mammals is characterized by playful behaviour and few animals are more playful than the otters.

tually, she does come fully on heat and now she submits, standing motionless for coition. This prolonged foreplay is not so much to reduce aggression between the pair, as in the courtship of birds, as to ensure that there will be an attendant male when full oestrus is reached.

Similar behaviour is encountered in cattle. The bull notices the cow two days before she comes in oestrus and stays near her. This is called the guarding phase and the bull often stands parallel to the cow, head to tail. As oestrus approaches, the bull becomes increasingly excited. He follows the cow closely and smells her, then performs flehmen. Other cattle are chased away and the bull paws the ground, or carves it with his horns, throwing earth over his back. Now he seriously attempts to court the cow by pushing and leaning against her, with his neck lying on her rump. If unreceptive she moves away but when in full oestrus she stands her ground, even pushes back as the bull mounts her.

Oestrus generally lasts only a few days (usually less than one day in cattle), but while it lasts sexual activity is intense. An exception is the bats in which, in northern latitudes at least, mating takes place at any time

BELOW: 'Mad as a March Hare' describes the spring behaviour of hares when they become excessively ebullient, chasing over fields and hillsides. The activity stimulates the females and, on occasion, males rear up and box with their forepaws.

Grey kangaroos dispute dominance. Arm waving is a sign that kicking with powerful hindlegs will soon follow.

African buffalo displaying the 'flehmen' grimace, rolling up its lips to show its gums. The significance of flehmen is not fully understood but it has something to do with detection of the female's odour and may also be used as a greeting.

between September and May and ovulation starts about May. A female bat impregnated in September can store and nourish sperms until she ovulates about 8 months later.

Copulation may be brief, just a few seconds in hoofed animals, or lengthy, sometimes lasting for many hours in carnivores. Where copulation is brief it may be repeated very frequently. George Schaller has recorded the intimate lives of lions and found that copulation lasts 6–68 seconds. During mating the male often grabs the lioness's neck, sometimes drawing blood. The lioness growls continuously and the male sometimes miaows. One male was watched for 2½ days. At the end of 24 hours he had copulated 86 times with two lionesses. In the next 24 hours he copulated a further 62 times and at the end of 55 hours' watching time his score was 157. The average time between mating was 21 minutes and was once as little as one minute. The record for copulation is, however, held by Shaw's jird, a desert rodent that has been recorded by Boulière as copulating an astonishing 224 times in two hours.

Mounting dorso-ventrally, or *more canum*, is the rule among mammals, even among bats which mate while suspended from walls or rafters. A slight variation is introduced by camels in which both sexes sit back like dogs. Belly-to-belly coupling, *more humanum*, is rare but is practised by Two-toed sloths, hamsters and whales. Mating in whales has been seen rarely but the pair couple very briefly side by side or poised vertically with heads clear of the water after a prolonged foreplay of slapping and caressing with the flippers. The mating habits of hedgehogs are the butt of risqué jokes and enquiries as to how they manage despite the coat of prickles. The answer is what might be expected: they mate as normal but take care. After circling the sow for some time, the boar hedgehog mounts her but runs little risk of being impaled as the sow flattens her prickles. Owners of pet hedgehogs will know that it is quite possible to stroke them when the prickles are flattened.

Boss males score. Lions are unique among the cat family for being sociable. The social unit is the pride, a stable organization of lionesses and their young. The one or two male lions in the pride are only loosely attached. They are replaced at intervals by others who have been leading a nomadic life and, when displaced, they return to wandering. Thus, at any time, only a few males have sexual access to the lionesses and, except in a few cases, this reflects the usual situation among mammals, and a fair proportion of males die without ever mating.

The Domestic cat is solitary but has a social system which limits the number of sexually active males. Each cat occupies a home range of paths, sleeping places and hunting grounds and, although neighbouring ranges overlap, the cats generally avoid each other; but they

Lions are social animals, unlike other cats. They live in prides – groups of lionesses, their cubs and two to four males. Courtship is playful but copulation is brief. The lion grasps the lioness's neck with his teeth and may draw blood.

often have friendly social gatherings and there is an order of precedence among males which is first decided by fighting and then maintained by displays. The product of the order of rank is a small clique of toms that have access to the females. The younger toms have to wait their turn until their seniors die or can be displaced.

A hierarchy of social standing is also the situation among mice. Mice live in groups which have a communal range or territory under the control of a dominant or 'boss' male. He mates with the females and keeps the other males submissive. Apart from physical fighting, the other males are kept submissive by pheromones from the dominant mouse. Mice, as well as cats and lions and many other mammals, deposit scent marks at intervals around their range. The marks are usually in the form of urine but may be scented droppings, as in rabbits, or the secretions from special glands, as in deer. Whatever their form, they act like the songs and displays of birds. They inform other animals of the sex and status of the animal that deposits them.

The pheromone communication systems of mammals have been studied only in the last few years but it is becoming clear that a pheromone has a far-reaching effect on the behaviour and physiology of the receiving animal. The senior mouse literally dominates the other males in the group because the pheromones passed in his urine prevent them coming into breeding condition. Their testes remain small and they remain juvenile. They

Male cubs stay with the pride until at least $3\frac{1}{2}$ years old, when they are forced to leave. They then live apart until they can fight their way back into a pride.

A charming sight: Prairie dogs kiss when they meet. If they are members of the same social group, they groom each other. If not, a fight breaks out.

can only mature if the dominant male dies or if they emigrate to find a place of their own.

The male mouse's pheromones also serve to attract female mice and subtly affect their breeding. If the dominant male is displaced by another, any of the females that were mated within the previous few days lose their newly developing embryos and come on heat again. This seems to be a piece of refined reproductive strategy. The new dominant male replaces the progeny of the ousted male in the females' bodies with his own so that he can contribute to the next generation as quickly as possible.

Harems and territories. Groups of mice and other rodents live in their territories throughout the year whereas many other mammals have different social organizations inside and outside the mating season. This is particularly noticeable among the hoofed animals where out-of-breeding season herds break up at the onset of the rut.

While birds proclaim their territory by singing, mammals usually deposit scent marking. This oribi, an African antelope, skilfully transfers scent from the facial gland to the tip of a grass stem.

The impala's rut starts when rains bring new life to the African savannahs. Males set up territories, ejecting challengers, and wait for the herds of females to pass by.

Saiga, the strange bulbous-nosed antelopes of the Russian steppes, live in mixed-sex herds until the mating season, when the mature males cut out small parties of cows from the main herd and defend them against other males. A single male may command as many as 50 females. The whole process of forming a harem and mating lasts only 7–10 days so that, unless properly timed, a visit to the saiga's home can easily miss the rut. During this period bull saiga develop an even more bulbous snout, along with woolly 'sideburns' and a yellowish chest, and they fight for dominance, sometimes maiming one another. As this takes place in December, the exhaustion from fighting and mating leads to the death of many males in the severe winter weather.

The Uganda kob, another antelope, has an arena style of mating like that of Black grouse. The males defend small, close-packed territories into which the females make their way to be mated. The position of the arena is immediately identifiable on the savannah grassland because the herbage is close-cropped and trampled over an area of about 200 yards diameter. Within the arena there may be 10–15 small territories but there are peripheral territories outside that are larger and scattered over the terrain. When they come on heat the female kob make their way to the central arena ignoring the males in the peripheral territories. As a female approaches a male kob in his territory, he prances towards her with head held high and the black hair of the forelegs and the white on the throat raised to make a showy display. The female stops and urinates. The male sniffs the urine, performs flehmen and touches her underparts or strokes her hindleg with a

RIGHT: The Red deer stag sounds his challenge. During the autumn rut the hills echo to the roars of stags as they threaten other stags attempting to steal the harem of hinds.

An imperious Fallow buck rounds up his harem so that he can defend them from challengers.

stiffly raised foreleg – an action known by its German name of laufschlag. Mating takes place immediately afterwards. It is brief and is followed by more displaying. Finally, the pair may feed or lie down together.

The female kob's preference for the arena leads to rivalry for the central territories. Competition is based on a stereotyped aggressive display of lowered head and ears turned sideways. Fighting is very rare but the continual activity of threatening and mating wears out the males very quickly and they cannot hold a central territory for more than a day or two. There is, therefore, an intense pressure of sexual selection acting on the males. So rigorous is the competition that males have a shortened lifespan. Superficially, it would seem that males have an easy life compared with the demands of maternity made on the females. Nevertheless, the competition to achieve paternity is gruelling.

One reason for the short tenure of a central territory in a kob arena is that there is hardly any sustenance for the resident male. He is forced to give way eventually through hunger. A similar problem is faced by Fur seal bulls. Breeding takes place on beaches where the bulls establish and maintain territories for long periods. The mating season is protracted because it coincides with the long birth season. Over a period of three months cows come onto the beaches, bear their pups and suckle them for six months. A week after giving birth each cow comes on heat and is mated. The bulls start to establish territories before the cows appear and they have to defend them until the cows are ready to mate. They cannot leave their territories without losing their place, so they are unable to eat. This is not too serious as they can survive on the thick layer of blubber, but the continuous exertion of defending the territory and of mating, wears them out and there is a succession of males on the breeding beaches.

The aim of every bull is to have a territory when the most cows are coming on heat and to

Young zebra mares are recognized by a special stance when on heat. This attracts stallions which abduct them from the family herd to lead a new life.

The rutting season of hoofed mammals is a time of intense sexual activity but activity is limited to a few males out of the entire stock. Competition results in polygamy, with the dominant males acquiring a harem and producing many offspring. Lesser males remain celibate.

be positioned where they bear their pups. Here the situation differs from that of the kob. The cows do not come seeking a male but haul out on land in the best places for bearing their pups and are fertilized with few preliminaries by the bull whose territory they are in. If a bull is on a stretch of flat beach he has a good chance of having a fair number of cows coming into oestrus in his territory, while lesser males must be content with areas behind the beach, along the tide's edge or on rocks. The bull tries to corral the cows on his territory and any that pass through are subjected to his attempts at herding. These are largely unsuccessful, hence the necessity to stake a territory where cows naturally gather.

Fur seal bulls are polygynous; only a fraction of the males obtain 'good' territories and only these few contribute to the next generation. The characteristics that allow them to be sexually successful are those that allow them to defend territories and stay on shore for a long time. The time a male spends on land depends on his ability to subsist without feeding and, for the closely related Californian sealion of warm Pacific coasts, territories must be along the water's edge or there must be pools where the bulls can cool off. To defend territories successfully, size and vigour are needed and there has been sexual selection in the polygynous Fur seals, sealions and true seals to produce males as much as four times the bulk of females.

Confrontation by beachmasters. Two adult male Elephant seals challenge each other at the territory boundary. The nose is inflated and the tip hangs over the mouth to act as a resonator for loud bellows of defiance. Around the bulls are grouped the cows which are ignored until they become receptive a few weeks after the pups have been born.

The Fur seals and sealions are relatively agile on land because they can tuck their hindflippers under the body and bound on all fours. The true seals cannot do this and sexual selection in Elephant and Grey seals has led to huge cumbersome males whose progress overland is reminiscent of a bloated caterpillar. They can move at a surprising speed but only over a very short distance. However, for the Grey seal of the North Atlantic, lack of mobility on land is probably a fairly new problem.

A few thousand years ago Grey seals bred on the frozen seas of the Ice Age, as they still do in the Baltic Sea. In recent times, Grey seals elsewhere have bred on beaches of remote islands. The cows drop their pups above the tide mark but spend much of their time, when not suckling, swimming offshore. The bulls establish their territories along the shore among the cows and, being expert swimmers, can defend their territories easily. In British waters at least, this situation has changed within the last century. The Grey seal is no longer the quarry of coastal people who hunted it for the sake of its pelt and

Courtship of elephants involves caressing and intertwining of the trunks.

blubber. Consequently its numbers have increased, the beaches have filled and cows have taken to pupping on land, behind the beaches. Their behaviour has changed to fit this new situation. The cows spend much, if not all their time with their pups until these are weaned and the bulls establish themselves among the cows. They now have the problem of reduced mobility and, to overcome this, Grey seal bulls have evolved a reproductive strategy different from that of Fur seals.

The cumbersome Grey seal bull cannot properly defend the boundaries of a land-bound territory and does not bother to do so unless an interloper tries to move in. Instead, he seems to co-exist peacefully with his neighbours and save energy so that he can stay on land longer. The biggest bulls stay ashore for as many as eight weeks, most of which are spent in sleep. Waking hours are devoted largely to sexual matters. The bull Fur seal investigates the sexual condition of his cows by smell, and mates with those in oestrus, but the bull Grey seal has a different approach. As he does not defend his territory, he cannot ensure his exclusive access to the cows.

the young bucks arrive, the doe will have been fertilized by the territory holder. The function of the young bucks is to provide a back-up service for does living with infertile bucks.

A closer family bond is maintained by foxes where dog and vixen stay together for most of the year and, as gestation is much shorter than in deer, the dog fathers the cubs of that year. The European Red fox is very closely related to the North American Red fox from which the Silver fox of the furrier has been bred. Family life starts in January and February after a solitary winter. The vixen marks her trails with scented urine to attract the dogs which roam the countryside in search of a mate. At this time of year, foxes are often heard calling at night. There is a yipping bark and the weird, harsh scream, usually described as the vixen's but sometimes uttered by the dog fox. The purpose of these calls is not clear although they may help the foxes find each other.

Compared with the aggression we have seen in the courtship of other animals, the pre-mating behaviour of foxes appears a very friendly affair. The pair play together in the same way as the cubs will play together in months to come. They romp together, chasing about and tumbling, and often rear up face-to-face with forepaws on each other's shoulders. While so poised, they lunge at each other with wide-open mouths, as if trying to bite each other's tongues. Baring the teeth with the mouth open is a sign of excitement and pleasure. Another posture comes at the end of the romp. The pair stop still, the dog standing in front of the vixen. He then throws his brush, or tail, over her shoulders, perhaps to bring his dorsal scent glands near the vixen's nose and so stimulate her.

Copulation proceeds as in dogs, where playing and running together are part of the

LEFT: Black rhinoceroses are rather solitary animals. The females attract males by their scent and by calling and one male may stay with a female for several days. At first she repels him but later she solicits his attentions.

BELOW: The antlers of the roe buck are still in velvet but he is marking his territory with scent from his facial gland.

Female fox standing in front of male with her tail turned to one side.

LEFT: Foxes frequently play together, rearing up to place forepaws on each other's shoulders or playing tag. The games are accompanied by a gaping grimace with the ears turned back as the foxes lunge at each other.

mating sequence, but the female does not accept the male until oestrus is well advanced and ovulation has taken place. At this time she will even invite copulation by standing in front of the dog with her tail turned to one side. After mounting, the dog's penis swells, locking the dog to the bitch in the 'tie'. When this is achieved the dog dismounts and turns round so that the pair are firmly locked rump to rump. Ejaculation takes place during the tie which lasts 10–30 minutes. This tie acts as a kind of 'guarding phase' to prevent other dogs gaining access to the female before it is fairly certain that the sperm of that particular dog will have fertilized the ova. In the promiscuous Domestic dog this is necessary, although it would seem to be superfluous in the monogamous foxes.

After a gestation of 50–56 days, the cubs are born and the dog fox's behaviour changes. Until then he has been completely selfish about food but he now brings prey to the vixen who devotes her time to nursing the cubs in the earth. When the cubs come above ground for the first time at 4–5 weeks, the dog's behaviour changes again and he holds the food in his mouth and encourages the cubs to leap up and snatch it, so laying the foundations of their hunting behaviour.

Sex appeal. The courtship play of foxes is charming to watch and the observer finds himself wondering whether the behaviour is more than a prosaic stimulation of each animal to make it ready for copulation. Do they have any finer feeling for each other and does either make a selection of mate, preferring one and rejecting others? We have seen in Chapter Seven how there are individual male Black grouse which have something that makes them more attractive than their fellows. In Black grouse, the 'something' is an increased aggression by the cock towards other males and an ability to match his courtship to the reaction of the hen. He has a 'style' that brings him success in his courting.

To find whether preferences occur requires very detailed observations. These are lacking for many animals but they are most easily carried out on domestic species. It is reported that there is no preference for partners in Domestic cattle. The bull mates with any cow in oestrus, provided she is not being guarded by another bull, and the cow accepts any bull provided he is socially dominant to her. So say the textbooks, anyway, but there is the story of the farmer whose pedigree cow was served by a pedigree bull three times and on each occasion she produced a stillborn calf. Yet when allowed to mate with a bull whom she had the habit of nuzzling across a fence, she attained motherhood every time. It seems that she had a definite preference for this particular bull which even influenced the mechanics of reproduction.

It is difficult for a single anecdote to challenge serious scientific observations but the idea of sexual preferences is supported by observations on dogs. Their behaviour is a byword for promiscuity but controlled tests on beagles show that bitches on heat accepted some dogs readily but rejected others totally, by growling and snapping at them. Furthermore, each bitch had a different order of preference for the dogs. What is more surprising, when not on heat, the bitches prefer to associate or play with certain dogs which may not be the ones they choose to mate with. A dog who is a good playmate is not necessarily a good sexual partner. The difference is probably due to the change in behaviour of the dogs when the bitches come on heat. From playing and other social activities they switch to courtship and the ardour and expertize of a dog's courtship may determine his sexual success.

Both mare and stallion become excitable during the rut and the mare allows herself to be chased. When allowed to run free horses form a stable society of one stallion with several mares but the stallion will not mate with his won offspring.

MONKEYS, APES AND MAN

Monkeys and apes belong to the order Primates. Man also is a primate. These facts have given a great stimulus to the study of monkeys and apes because, as our closest relatives, it is thought that they may tell us something about ourselves. It is also possible to carry out experiments on monkeys and apes that would be unthinkable if Man were the subject. Consequently, scientists from every field of zoology have descended on the non-human primates and, in the field of behaviour, sociologists, psychologists and anthropologists have turned to them in the hope of shedding light on the most fascinating of all animals – Man himself. The zoologists, who study animals other than Man, are more interested in the study of primates for their own sake. Their observations should provide a warning to anyone who attempts to understand Man by drawing too close comparisons with his remote ancestors. There is a very wide variety of primates living throughout the warmer parts of the world. In appearance they are very different from each other and so they are in their behaviour. Man is no exception to this. Indeed, he is a very unusual primate and any apparent similarities between the behaviour of this or that monkey or ape and Man must be treated very cautiously.

This point must be stressed. As the study of non-human primate behaviour has progressed, it is found that, even within a single species, social behaviour can vary quite considerably. On the other hand, two quite distantly related species may have very similar behaviour. In both instances, the factor controlling behaviour is, not surprisingly perhaps, geared to the way of life and physical milieu or environment of an animal and, as Man has adopted many ways of life, his social behaviour is similarly varied.

With these reservations in mind, comparisons of Man and his close relatives are, to say the least, amusing, as any visit to the Monkey House shows. They may possibly also be of great importance in understanding the basis of our social behaviour. Such an idea has reached the popular level in Desmond Morris's *The Naked Ape* which examines our sexual behaviour and in Robert Ardrey's *Territorial Imperative* which seeks to explain our aggression towards members of our own kind. Both books have enjoyed considerable popularity and deservedly so. There have, however, been grumbling reservations from the experts who are worried by too direct a comparison between Man and other primates. The value of studying primate behaviour lies in its simplicity. The basic forces and pressures acting on an individual or a society of monkeys or apes can be seen more clearly than those acting on Man. However, once these factors are defined it is possible to investigate their role in human society. The criticism of direct comparison is made the more valid because Man is such a variable, complex being that it is difficult to find a common denominator of human behaviour which can be used in such comparisons. To solve this problem, considerable research and speculation have been made about the social life of primaeval Man when he first evolved from a man-like ape to an ape-like man and before he acquired a patina of civilization.

There are some important trends in the evolution of the primates which have affected their social lives. The early primates were insect-eaters, like many of the lemurs of today, whereas most of their descendants have become plant-eaters. The early primates had long snouts and depended on a good sense of smell. They scampered through the trees in search of insects. The monkeys took to swinging from branch to branch, for which they needed good eyesight. Thus we find the

Chimpanzee mating is quick and casual.

long snout replaced by a flattened face with forward directed eyes, and the replacement of smell by vision as the dominant sense has affected the courtship of primates. There has been a shift in communication between male and female from the use of pheromones to visual signals. The shift is not complete, however. Pheromones still play a part in the mating of the higher primates and there is intriguing evidence that the sense of smell may be used in the courtship of human beings.

Because there is such a range of differences between the many and various primates, it is not easy to present a simple picture of their mating systems; except to say that courtship itself is uncomplicated. Female monkeys and apes have a menstrual cycle as in Man but with the important difference that they are only receptive to the male during a period of oestrus. The onset of oestrus is signalled by visual or olfactory signs. In some species there is a grotesque enlarging and colouring of the external genitalia such as the enormous pink swelling of the female chimpanzee which is stimulating to a male chimpanzee but revolting to our eyes. The Rhesus monkey female secretes a pheromone from the vagina. Called copulin, it is essential for arousing the interest of males. So, courtship appears crude and casual to our eyes: the male inspects the female's genitalia by sight or smell and, if the signal is right, mounts her. This perfunctory courtship would seem to go against all that has been said about the need for courtship displays to ensure selection of the correct mate. The apparent paradox is resolved by the pair already being familiar with each other. Unlike Roe deer, Uganda kob, albatrosses or grouse, primates live in permanent, year-round societies. Most members of a troop will have known each other from birth and know each other individually. They are well aware of the sex, social status and character of each of their fellows. Consequently, there is no need for elaborate display or even territorial behaviour.

Rigid baboon societies. Primates live in stable societies which may number scores of individuals, as in baboon troops, but, at the other end of the scale, gibbons and the orang-utan of dense forests live in small groups. Gibbons are monogamous and live in families of two adults and their young. The stability of primate society is illustrated by the baboons – long-muzzled, African monkeys that have left the comparative safety of the trees for open country. As is usual with primates that live on the ground, the males are much larger than the females. Their size, backed by large canine teeth, is a valuable defence against leopards and lions.

The baboons of the African savannahs live in troops of 20–80 individuals. The society of the troop is tightly knit. Only rarely does a baboon change troops, and movement is mainly of mature males who leave one troop, perhaps because their advancement up the

RIGHT: As with preening in birds, mutual grooming is employed by mammals as a means of binding social relationships.

scale of rank is blocked by too many dominant males. When it enters its new troop, the baboon has to fight for a place in a hierarchy so rigid that the rank of a baboon affects its physical position as the troop moves across open ground. There is a definite order of march. The dominant males accompany the females with infants in the centre and the weaker males and young animals are spread around the periphery. Here, they are in a position to give warning of attack and act as a barrier to safeguard the vulnerable infants and their mothers. When danger threatens, the dominant males move forward to investigate and, if necessary, attack.

A female baboon comes into oestrus and is receptive for one week in every month. As oestrus approaches, she leaves the small group of 'friends' with whom she normally associates and even abandons her offspring temporarily, to join the males at the edge of the troop. At first, the dominant males are not interested and she mates with the lower orders. She may solicit copulation with a dominant male during this period by adopting the mating posture in front of him but she is ignored until she comes fully into oestrus and is fertile. Then, a dominant male assumes control of her and they form a 'consort pair'. The pair withdraw to the edge of the troop and spend their time together. Other males are driven away and the male consort is assured of fertilizing the female.

An unexpected change of behaviour comes over the baboons in a consort pair. Normally a baboon is selfish about food and strong animals steal from the weaker but an experiment with captive baboons shows that the consort pair learn to co-operate. A male was denied access to a plate of food except by pulling it through a small gap in his cage with the aid of a hooked stick. After he had learned to do this, the stick was also placed out of reach. The poor male was frustrated until he formed a consort pair, when his new mate suddenly perceived the problem and its solution. She brought him the stick and, thereafter, the two co-operated regularly to get food. Moreover, the female's intelligence and co-operation were rewarded because the male became noticeably generous in sharing his food with her. Other baboons, of both sexes, were present in the cage but none showed any signs of co-operation. What we would like to know now is whether such behaviour is a

Verreaux's sifaka is one of the lemurs that lives in the forests of Madagascar. Sifakas live in small troops and the females come into season together and are mated by the males.

The capuchins are the best known of all American monkeys although little is known of their behaviour in the wild. They live in troops in which adult males keep guard over females with infants.

Male Rhesus monkeys groom their females when they are in season; otherwise it is the females that do the grooming. A dominant male allows no other males near the females.

normal occurrence in the wild and, if so, whether the female baboon derives any real and permanent benefit.

Promiscuity in chimpanzees and gorillas. The problem of studying ape and monkey behaviour in the wild is that their forest-dwelling habits often make them very difficult to watch. Virtually nothing is known of the mating habits of the orang-utan and the gibbons, and in *The Year of the Gorilla* George Schaller was able to describe copulation only twice. Gorillas live in troops consisting of one male and one or more females with their young. Life is peaceable and solitary males may join the troop for short periods. The resident male even allows them to mate with the females. Schaller watched a female mate with such a male. She took the initiative by embracing his waist and rubbing against him. He did not require much enticement and quickly pulled her into his lap. The resident male came towards them and the strange male broke away and retreated. He only returned after the resident had wandered away and the union was then duly consummated.

The chimpanzee has proved more amenable to study than most primates and Jane Goodall's lengthy study of wild chimpanzees in the Gombe Stream Reserve has given us a good picture of their love life. The chimpanzees live in small groups, the composition of each being fluid, as individuals are free to come and go. There is a rank system, in which subordinate animals defer to their superiors but this does not affect mating behaviour. Chimpanzees are promiscuous, there being no sexual selection or restriction through low social status or lengthy pairing-off. Occasionally, a sort of consort pair is formed but the male does not seem to object to other males copulating with his female. More usually a receptive female with the characteristic pink swelling is followed by a number of

Baboons are monkeys which have taken to living on the ground. Here a female presents herself to a male and invites his attentions. While she is on heat the female will spend her time with one particular male who chases others away.

Baboon society consists of a rigid hierarchy in which males work out a social scale by fighting.

males who mate with her in turn. Indeed, one chimpanzee was seen to mate with eight different males within the space of 15 minutes. In such circumstances it must be a matter of chance which male contributed the sperm that eventually reached the ovum.

Nevertheless, in the long run there is some sign of sexual selection in chimpanzees. The males are attracted particularly to fully adult females, and young females may wander alone even when they have sexual swellings. The majority of matings involve mature males because the females are more responsive to their advances and young males are particularly wary of copulating when there are dominant males nearby. Jane Goodall saw a female solicit a young male who glanced anxiously at two nearby adult males before moving to a spot 10 ft away, where mating took place.

Baboons, gorillas and chimpanzees have very different social systems. Indeed, every type of primate has its unique system. Coupled with the fact that many species have not been well studied, this means that it is impossible to summarize the mating strategies and behaviour of primates in a succinct fashion. Moreover, the diversity of primate behaviour is increased enormously when we realize that Man himself, with all his different cultures, is also a primate and must be included in any discussion.

What about Man? It is fitting to finish a discussion of the mating of animals by speculating on how Man fits into the general picture of courtship displays and repro-